INTERNATIONAL CENTRE FOR MECHANICAL SCIENCES

COURSES AND LECTURES - No. 121

E. R. de ARANTES e OLIVEIRA
UNIVERSITY OF LISBON

FOUNDATIONS OF THE MATHEMATICAL THEORY OF STRUCTURES

SPRINGER-VERLAG WIEN GMBH

Originally published by Springer-Verlag Wien New York in 1975

ISBN 978-3-211-81312-6 ISBN 978-3-7091-4328-5 (eBook)
DOI 10.1007/978-3-7091-4328-5

1. INTRODUCTION

The differential equations of Mathematical Physics are very often associated to variational principles which state that the solution of the corresponding equation, under given boundary conditions, makes a certain functional stationary on a certain space of functions.

Those methods which replace the problem of solving the equation by the equivalent problem of seeking the function which makes the functional stationary are called variational methods.

A classical variational technique is the Ritz method, which reduces the problem of the minimization of a given functional F on a given space C to the minimization of the same functional F on a finite-dimensional subspace C' of C.

The finite element method is also a variational method in which the elements of C' are piecewise defined on a given domain. It does not always coincide with the Ritz method, however, because the finite-dimensional set C', on which F is made stationary, is generally not contained in C.

It became thus necessary to generalize the old theory to cover the new situations. New convergence theorems were namely stated and demonstrated. The presentation of such theory, using the concepts of functional analysis is the aim of Chapter 3.

The theory supposes that an extremum principle exists. Other variational principles (even Galerkin's method) can be used for establishing sequences of approximations, but the convergence of such sequences cannot be proved, within the frame of the present theory, without the help of an extremum principle.

Although the new convergence theorems were established with the aim of being applied to the finite element method, the theory has a much more general scope. Structural and non-structural applications can indeed be considered and even the linear assumption is not necessary.

Although non-structural applications can be covered, one of the most interesting applications of the convergence theorems appears in the theory of structures as it is shown in Chapter 4.

Chapters 5 and 6 respectively introduce the three- and two-dimensional models of the theory of structures. The discrete model is finally considered in Chapter 7.

A short account of the evolution of the finite element theory and of the papers which the author has been publishing on the subject will be presented now.

The finite element method is a discretization technique for the solution of differential equations, the characteristic feature of which is the superposition of coordinate fields piecewise defined on the domain.

Three stages canbe distinguished in the method:

a) subdivision of the domain into subdomains;
b) definition of the finite-dimensional set of fields allowed within each domain (discretization of the field within each subdomain);
c) definition of the interaction between the different subdomains, i.e. of the way in which the field within a given subdomain is connected with the fields of the contacting subdomains.

Different modalities can be considered according as the allowed field within and the interaction between the elements are defined.

In the oldest and most used modality, presented by Turner, Clough, Martin and Topp [1] in 1956, the allowed fields are defined through the displacements, and the interaction through compatibility conditions.

The second modality (hybrid elements), in which the fields are defined through the displacements, and the interaction through compatibility conditions, is due to Pian [2] (1964).

A third modality was presented by de Veubeke [3], also in 1964, in which the fields are also defined through the stresses, but the interaction is defined through equilibrium conditions.

Finally a fourth modality (mixed elements), in which the fields are defined partly through the stresses, partly through the displacements and the interaction partly through equilibrium, partly through compatibility conditions, was introduced by Herrmann [4] in 1965.

In all the four modalities the problem arises of which criteria must be followed for the discretization of the field inside the subdomains.

Such problem was first discussed in connection with the first modality and became critical when the finite element technique started to be applied in the analysis of transversely loaded plates [5].

Up to that time it was assumed, more or less consciously, that continuity of the displacement across the element boundaries was a necessary and sufficient condition for the success of the method. Such condition had been satisfied without difficulty while problems of plane elasticity were the only ones considered. It ceased to be so with plates because, then, the generalized displacements are the transverse displacements and the rotations of the normal to the middle plane, and the continuity of the rotations across the element boundaries proved very difficult to achieve.

Certain unsuccessful results obtained at first were ascribed to the violation of continuity. Nevertheless, as soon as complete compatibility was secured with rectangular elements [6], it was seen that such elements yielded unsatisfactory results, which where explained by the fact that the sequences of approximate solutions generated by systems of elements with decreasing size did not converge to the exact solution.

The condition of the convergence to the exact solution appeared to be of fundamental importance, particularly after having been remarked that certain corrections applied to triangular elements, to ensure compatibility, really decreased the speed of convergence, which, after all, was what was important to increase [7].

A convergence criterion for the first modality was presented in 1965 [8]. According to it, every possible state of uniform strain must be allowed within the element.

In plane elasticity, such criterion requires that the polynomial expression for each displacement component contains an arbitrary constant and two linear terms multiplied by coefficients also arbitrary. Applied to the theory of plates and accepting Kirchhoff's assumption, which implies that rotations be derivatives of the transverse displacement, the criterion requires that the expression of the later contains at least an arbitrary constant and all the linear and quadratic terms also multiplied by arbitrary constants.

The above criterion was based on heuristic considerations and was not, at first, properly speaking demonstrated. A demonstration followed almost immediately for the cases in which compatibility is not violated. Nothing else was necessary than noticing that the finite element method became, in such case, a particular case of the Ritz method [14], and then applying to the finite element method the well-known convergence theory of the latter.

As it is known, given a certain functional F defined on a linear space, C, Ritz method makes it possible to determine the element of C which minimizes functional F. For this purpose, a sequence of linear subspaces of C with a finite number of dimensions is considered, and the element which minimizes F in each subspace is determined. The sequence of such elements converges to the exact solution, i.e. to the one which minimizes F in C, if the sequence of linear subspaces is complete with respect to a subset of C which contains the exact solution, i.e. if, given an element e whatsoever of the subset, it is possible to obtain a sequence converging to e made up of elements of the successive linear subspaces. Such is the so-called completeness criterion.

Now, in the case of the first modality, the functional to be minimized is the total potential energy and the linear space is the set of all the compatible fields. The successive subspaces are families of compatible fields generated by the successive systems of elements, the number of dimensions of each subspace being equal to the total number of nodal displacements corresponding to each system.

What was demonstrated was that the convergence criterion presented for the finite element method, which will be, from now on, more properly called completeness criterion, was no more that the completeness condition with respect to the set of compatible fields whose strains have bounded and continuous first order derivatives within each subdomain.

The case of the first modality for which compatibility was violated, remained unsolved, not to speak of the order modalities.

Before proceeding further, it should be noted that convergence and completeness have no meaning unless a definition for the distance between two fields is introduced. Indeed, a sequence of fields is said to converge to a limit when the sequence of distances between each of its terms and the limit tends to zero.

Convergence is uniform when the distance between two fields is defined in terms of the maximum modulus of the difference between the values of the fields at each point of the domain.

Now, the concept of uniform convergence usually applied in the finite difference method, proves poor when applied to a method like the finite element one closely connected with energy concepts (°).

On the other hand, it is much stronger than needed, as in practice it is not necessary that two fields coincide at all points, in the limit. In other words, uniform convergence may be replaced by energy convergence.

According to the precedent remarks, the square root of the strain energy of the difference between two fields was taken as the measure of the distance between them.

Such is the concept of distance used in Mikhlin's study on convergence in the Ritz method presented in his famous book [11] on variational methods in mathematical physics. This and other basic mathematical concepts used on some papers on finite elements were indeed supplied by Mikhlin, although the finite element method itself is mentioned nowhere in the book.

A paper [13] was published by the author in

(°) It is important to remember, however, that it was proved by Johnson and McLay, [12] in the case of the first modality without compatibility being violated, that, although stresses converge in the mean, displacements convergence uniformly for some kinds of elements.

1968 which is an attempt to describe the finite element method (first modality) as a general method for the solution of the very general class of elliptical equations considered by Mikhlin in his book. Such class comprises namely Navier's equations of two and three-dimensional elasticity and Lagrange's equations of the theory of plates. Following Mikhlin's example, the concepts and language of functional analysis were used in the description.

For such description, it became necessary to introduce the concept of principal derivatives, i.e. the derivatives of the field which must be kept continuous, in order that the energy remains bounded. In the case of Navier's equations of two and three-dimensional elasticity, principal derivatives have order zero, as only the actual displacements are to be kept continuous. In the case of Lagrange's equation, which appears in the theory of plates if Kirchhoff's simplification is introduced, principal derivatives have orders zero and one, as both the transverse displacements and their first derivatives must be continuous (°).

(°) However, if the transverse shear deformations are not neglected, principal derivatives are again of order zero, since the rotations of the normal to the middle surface are displacements just as the transverse displacement.

It was then demonstrated that the completeness criterion, with respect to the set of fields presenting continuous and bounded second derivatives of the principal derivatives, is that the first order derivatives of the principal derivatives of the field can assume arbitrary constant values within each element.

The same paper contains an analysis of convergence in the general case of the continuity of the principal derivatives being violated across the element boundaries. Notice that the discontinuity of the principal derivatives means violation of compatibility if the structural theories are concerned.

The basic concepts of this analysis had been developed first in connexion with a theory of shells [14, 15] involving moments with order up to N, in an attempt to prove that the solution yielded by this theory tends to the solution supplied by three-dimensional elasticity as N tends to infinity. This demonstration was adapted without much difficulty to the finite element theory, and it was readily understood then that a general convergence theorem, applying to the passage from a general structural model to another was implicit.

The analysis showed that completeness alone does not necessarily imply convergence if compatibility, or, in general terms, if the continuity of the principal derivatives is not achieved across the element boundaries. A supplementary condition must hold, viz. that the second derivatives of the principal derivatives remain bounded inside the elements when their size decreases indefinitely.

This supplementary condition was at first considered as disappointing, since completeness was then believed to be a sufficient condition for convergence, whether compatibility was violated or not.

Previous work carried out by Zienkiewicz's team [7] was certainly known which seemed to indicate that the plate element developed by this team did not always converge. It was thought however that the small convergence errors observed would be explained in any other way.

It became clear however, that such uncomfortable results were really due to deficiency of convergence, so that the cause was ascribed to the supplementary condition which, it was thought, could be dispensed with.

This led to a comment [16] on the paper in reference, prepared in collaboration with Zienkiewicz and Irons, in which it was explained that, in the case of plates, no convergence can be obtained unless the third derivatives of the transverse displacement remain bounded within the elements when their size decreases indefinitely. This condition is fulfilled when the elements are arranged in such a way that the nodes are all of the same kind.

Analogous considerations were contained in

a paper [17] presented at the second Dayton Conference on matrix methods where the analysis in reference was particularized to structural models.

The arrangement of the elements has no influence if compatibility is not violated across the element boundaries. On the other hand, in the case of plane or three-dimensional elasticity, the second derivatives of the principal derivatives (which are of order zero) are the second derivatives of the displacements, which remain bounded whenever the field within each element is equilibrated by body forces with bounded density. This is particularly the case if such density vanishes. Completeness then ensures convergence, even without compatibility between the elements being achieved.

At this point it was understood that it would be convenient to work at the level of the general theory of structures. Indeed, on one hand, the studies carried out on the finite element method suggested new bases for the synthetic formulation of the theory; on the other hand, it was hoped - correctly, as subsequent events confirmed - that an overall view could help to clarify some particular cases.

The scheme of the synthetic formulation of the theory of structures had already been approached by the author in 1966, [18] in an attempt to unify, by means of the variational theorems, the different continuous and discrete models or theories applied in the analysis of the different types of structures.

It was recognized however that a serious limitation of variational theorems is that they can supply no information on convergence if they are not minimum theorems. This is why the basic role passed to the convergence theorems, on the basis of which it was possible to formulate an acceptable mathematical theory of structures [19] .

2. BASIC CONCEPTS OF FUNCTIONAL ANALYSIS

2.1 Functions and Functionals

Let X and Y be arbitrary nonempty sets and suppose that some rule is given whereby to each element $x \in X$ corresponds a uniquely determined element $y \in Y$ which is called the image of x in Y. Consider the set F consisting of all ordered pairs (x, y) where $x \in X$ and y is the corresponding element of Y. Such set of pairs is called a function or an operator on X into Y [21]. The set X is called the domain of the function and the set of the elements of Y corresponding to all the elements of X is called the range of the function.

The set of all ordered pairs (x, y) with $x \in X$ and $y \in Y$ is called the cartesian product of X and Y, denoted by $X \times Y$. A function with domain X and range in Y is thus a particular subset of $X \times Y$.

Let F be a function with domain D and range R, where $D \subset X$ and $R \subset Y$. Consider the cartesian product $Y \times X$ and the subset $Y \times X$ consisting of those elements (y, x) such that $(x, y) \in F$. If such subset of $Y \times X$ is a function with domain R and range D, i.e., if the correspondence between $F(x)$ and x is a one-to-one correspondence between the elements of R and D, then we call it the inverse function to F, and we denote it by

F^{-1}.

If Y is the field of real numbers, then an operator on X into Y is called a functional on X.

2.2 Linear Spaces

A collection of elements together with a certain structure of relations between elements, or of rules of manipulation and combination, is called an abstract space [21].

A group is an abstract space, X, in which a law of combination called addition of any pair of elements, x and y, to yield another element denoted by $z = x + y$, is defined, such that

a) $x + y$ is an element of the space;

b) the space contains a unique element denoted by 0 and called the zero element, such that $x + 0 = 0 + x = x$;

c) $x + (y + z) = (x + y) + z$;

d) to each element x corresponds a unique element, denoted by $(-x)$, and called the inverse of x, such that $x + (-x) = 0$

If the addition is commutative, i.e. if $x + y = y + x$, the group is called an Abelian group.

An Abelian group is called a linear or a vector space, on a given field of numbers $\alpha, \beta, \ldots$, and its elements are called vectors if an operation called multiplication is defined, with respect to such field of numbers, such that

e) $\alpha(x+y) = \alpha x + \alpha y$;

f) $(\alpha+\beta)x = \alpha x + \beta x$;

g) $\alpha(\beta x) = (\alpha\beta)x$;

h) $1.x = x$;

i) $0.x = 0$.

A subset L of a linear space X is called a linear subspace, or a linear manifold of X, if all the elements of the form $\alpha x + \beta y$ belong to L, once x and y also belong to L. A subspace is called proper if it is not all of X.

The set consisting of 0 alone is thus a linear subspace of X. Supposing that S is any nonempty subset of X, the set M of all the linear combinations of the elements of S is also a linear subspace of X which is said to be generated, or determined, or spanned by S.

A linear space is called N-dimensional if it can be generated by a subset with N elements and no other subset with a lesser number of elements.

If a constant element is added to the elements of a linear subspace S of X, we obtain what we call a coset of S.

A set of vectors $x_1, \ldots x_m$ are linearly independent if $\sum_{i=1}^{N} \alpha_i x_i = 0$ implies $\alpha_1 = \alpha_2 = \ldots \alpha_N = 0$. Any linearly independent subset of X which generates the whole space is called a basis of X . It can be demonstrated that any two basis of a finite dimensional vector space have the same number of elements. Thus, the number of elements of any basis must be equal to the number of dimensions of the space.

Any vector x can be expressed in a unique way as a linear combination of the vectors x_i of the basis, $x = \sum_{i=1} \alpha_i x_i$, and the coefficients α_i are called co-ordinates of the vector x with respect to $x_i, x_2, \ldots, x_N$.

A linear operator A on X into Y is a function, whose domain is a linear space X and whose range is contained in another linear space Y , such that

$$A(x + y) = Ax + Ay \tag{2.2.1}$$

$$A(\alpha x) = \alpha A x \tag{2.2.2}$$

α being an arbitrary number and x and y arbitrary vectors of X .

Two linear spaces X and Y , with the same field of numbers are said to be isomorphic if there exists a linear operator T whose domain is X whose range is all of Y , and whose inverse T^{-1} exists. Two isomorphic spaces must have the same number of dimensions.

2.3 Metric Spaces

Let X be any set and consider the cartesian product $X \times X$ of X by itself. A distance function on X is a functional with domain $X \times X$ such that

a) $$d(x_1, x_2) = d(x_2, x_1);$$

b) $$d(x_1, x_2) = 0 \text{ if and only if } x_1 = x_2;$$

c) $$d(x_1, x_3) \leq d(x_1, x_2) + d(x_2 + x_3);$$

for arbitrary x_1, x_2, x_3. Inequality c) is called triangular inequality.

Making $x_1 = x_3$ in c) and using a) and b), there results

$$d(x_1, x_2) > 0$$

whenever x_1 and x_2 are distinct.

A set X with a distance function is called a metric space. Its elements are called points.

An element x of a metric space M is called limit (°) of a se-

---

(°) The concepts of limit and convergence are introduced with the help of the notion of distance. This is not necessary however as such concepts can be defined in more general structures which are called topological spaces. Such generality will not however be required in the sequel [21] .

quence of elements $x_1, x_2, \ldots, x_n, \ldots$ of M if $d(x_n, x) \to 0$ as $n \to \infty$. The sequence is then said to converge to x. Any sequence is called convergent whenever it admits a limit.
We call the set of all points $x \in X$ which satisfy the inequality $d(x,a) < r$ a sphere with centre a and radius r. A neighborhood of a point x is every sphere with x as centre. Any subset of X is said to be bounded if it is contained within some sphere.

Given a set $M \subset X$, a point $x \in X$ is called an accumulation point of M if every neighborhood of x contains at least one point of X distinct from x. The set obtained by adding to M every accumulation point of M is called the closure of the set M and is denoted by $\bar{M}$.

A set M is said to be closed if $M = \bar{M}$. A set is open if its complementary set, $X - M$, is closed. A set $M \subset X$ is called dense in another set $G \subset X$ if $M \subset G$ and $G \subset \bar{M}$. The space X is called separable if there exists a finite or countable set which is dense in X.

A sequence $\{x_n\}$ of elements of a metric space X is called a Cauchy sequence or a fundamental sequence if, given an arbitrary positive number ε, there exists a positive number N such that $d(x_m, x_n) < \varepsilon$ for $m > N$ and $n > N$. It is easy to prove that if a sequence x_n converges to a limit x, then it is a Cauchy sequence.

The converse of such theorem is not true for any metric space X. If it is, i.e., if any Cauchy sequence converges to a limit which belongs to X, then the space is called complete.

A function F on X into Y is said to be continuous at a point $x_0 \in X$ if to each neighborhood V of $f(x_0)$ in Y there corresponds a neighborhood V of x_0 in X such that $f(U) \subset V$.

Supposing that F is a function with domain X and range Y, such that the inverse function F^{-1} exists and that both F and F^{-1} are continuous on their domains, F is called a homeomorphism of X onto Y, and X and Y are said to be homeomorphic. Function F is also called a topological mapping of X onto Y.

Two metric spaces are said to be isometric if there is a function F with domain X and range Y such that the inverse function F^{-1} exists and furthermore that $d(x_1, x_2) = d(f(x_1), f(x_2))$ for every pair of points $x_1, x_2 \in X$.

2.4 Normed Spaces

A linear space E is called a normed linear space if to every element $x \in E$ a well defined real number is associated, called the norm of x and denoted by $\|x\|$, which respects the following axioms:

a) $\|x\| \geq 0$;

b) $\|x\| = 0$ if and only if $x = 0$;

c) $\|x + y\| \leq \|x\| + \|y\|$;

d) $\|\alpha x\| = |\alpha| \, \|x\|$;

α being any number.

It is easy to see that we can always introduce a metric in a normed linear space by making the distance $d(x,y)$ between two points x and y equal to the norm of their difference

$$d(x, y) = \| x - y \| \tag{2.4.1}$$

and, after the metric is introduced, we can define convergence as before. This kind of convergence is called norm convergence or strong convergence.

A complete normed linear space is called a Banach space. An operator A is bounded if

$$\| A x \| \leq C \| x \| \tag{2.4.2}$$

C denoting a positive constant. The least of the values of C satisfying (2.4.2) is called the norm of the operator and is denoted by $\| A \|$.

It is possible to demonstrate that any linear continuous operator is bounded and that any linear bounded operator is continuous.

2.5 Hilbert Spaces

A real (°) linear space X is called a real inner-product or a Hilbert (°°) space [11] if there is defined on $X \times X$ a real-valued function (x_1, x_2) called the inner product of x_1 and x_2 with the following properties

a) $$(x_1, x_2) = (x_2, x_1)$$

b) $$(x_1 + x_2, x_3) = (x_1, x_3) + (x_2, x_3)$$

c) $$(\alpha x_1, x_2) = \alpha (x_1, x_2)$$

d) $$(x, x) \geq 0$$

e) $$(x, x) = 0 \text{ if and only if } x = 0.$$

Axiom d) implies Schwartz's (°°°) inequality:

$$(x_1, x_2) \leq \sqrt{(x_1, x_1)(x_2, x_2)} \tag{2.5.1}$$

(°) The definition can be extended to complex linear spaces and complex-valued inner-products.

(°°) Some authors call Hilbert space only an infinite dimensional inner product space. If the number of dimensions is finite they call it an Euclidean space [21, 22] .

(°°°) Others call it the Cauchy's inequality.

which can be easily demonstrated by considering the inner-product of the vector $\alpha x_1 + x_2$ by itself

$$(\alpha x_1 + x_2, \alpha x_1 + x_2) = \alpha^2 (x_1, x_1) + 2\alpha (x_1, x_2) + (x_2, x_2) \tag{2.5.2}$$

In order that the quadratic expression of the right-hand side of (2.5.2) be non-negative, for arbitrary x_1 and x_2, its discriminant must be non-positive, and thus

$$(x_1, x_2)^2 - (x_1, x_1)(x_2, x_2) \leq 0 \tag{2.5.3}$$

which implies (2.5.1).

It is always possible to define norm and distance in a Hilbert space by making

$$\|x\| = +\sqrt{(x, x)} \tag{2.5.4}$$

$$d(x_1, x_2) = \sqrt{(x_1 - x_2, x_1 - x_2)} \tag{2.5.5}$$

Axioms a) and b) and d) of (2.1.4) are indeed satisfied. In what concerns axiom c), it suffices to remark that by virtue of Schwartz's inequality

$$(\|x\| + \|y\|)^2 = \|x\|^2 + \|y\|^2 + 2\|x\|\,\|y\| = (x, x) + (y, y) +$$
$$+ 2\sqrt{(x, x)(y, y)} \geq (x, x) + (y, y) + 2(x, y) = (x + y, x + y) = \|x + y\|^2 \tag{2.5.6}$$

2.6 Orthogonality

Two elements x and y of a Hilbert space H are said to be orthogonal if their inner product equals zero, i.e. if

(2.6.1) $$(x, y) = 0$$

If two elements of H are orthogonal to x, then every linear combination of such elements is also orthogonal to x. This means that the set S of all the elements of H orthogonal to x is a linear subspace of H, the linear subspace orthogonal to x.

A subset $D \subset H$ is said to be orthogonal to x if, for any pair of points $y_1, y_2 \in D$,

(2.6.2) $$(x, y_1 - y_2) = 0$$

The union of all subsets of H which are orthogonal to x and contain a given element, e, is a coset of the linear subspace orthogonal to x.

Two subsets C and D are said to be orthogonal if, for any pair of elements $x_1, x_2 \in C$ and any pair of elements $y_1, y_2 \in D$,

(2.6.3) $$(x_1 - x_2, y_1 - y_2) = 0$$

In order that two subsets can be orthogonal they must be contained in cosets of linear subspaces of H.

The linear subspace $R \subset H$ is the orthogonal complement of a linear subspace $S \subset H$ if it is the set of all the elements of H orthogonal to S. The whole space H is then called the orthogonal sum of R and S.

Two orthogonal subsets C and D cannot intersect in more than one point. Assume indeed they intersect in two points, x_1 and x_2. As x_1 and x both belong simultaneously to Cand D, we have by virtue of (2.6.3)

$$(x_1 - x_2, x_1 - x_2) = 0 \quad (2.6.4)$$

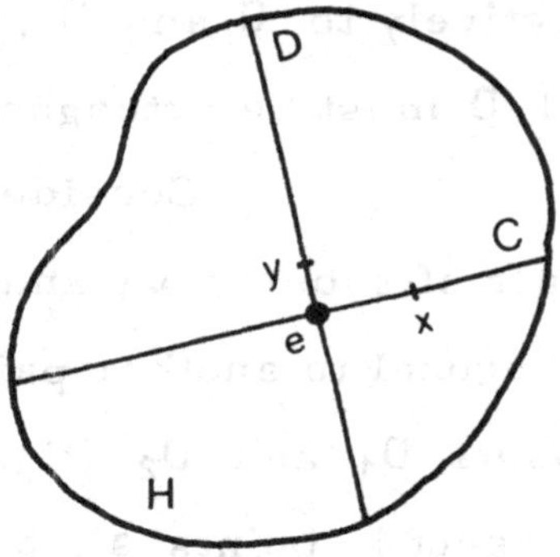

Fig. 2.1

and this can be true only if x_1 coincides with x_2.

Pythagoras theorem is valid in Hilbert spaces. Indeed, if the subsets C and D are orthogonal to each other (fig. 2.1) and if $x \in D$ and $y \in D$, then

$$[d(x,e)]^2 + [d(y,e)]^2 = [d(x,y)]^2 \quad (2.6.5)$$

e denoting the intersection of C and D.

Indeed, by virtue of (2.5.5) and (2.6.3),

$$[d(x,e)]^2 + [d(y,e)]^2 = (x-e, x-e) + (y-e, y-e) =$$

$$= (x-e, x-e) - 2(x-e, y-e) + (y-e, y-e) =$$

(2.6.6)

$$= (x-e-y+e, x-e-y+e) = (x-y, x-y)$$

The distance between x and any point $y \in D$ takes thus its minimum value if $y \equiv e$. Such value is called the distance between x and D .

Conversely, if (2.6.5) holds for any elements x and y belonging respectively to C and D , then, C and D must be orthogonal.

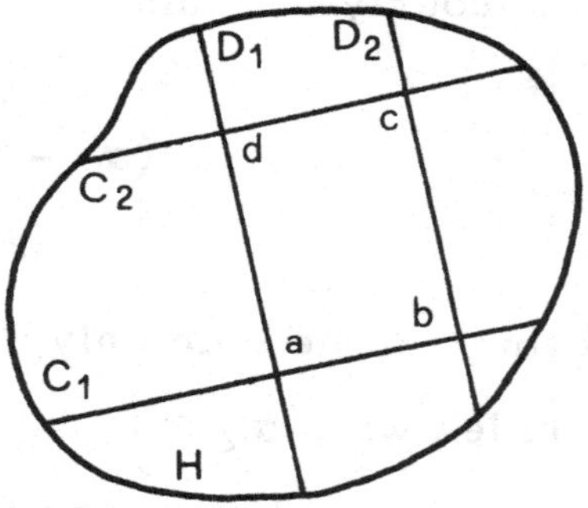

Fig. 2.2

Consider now a pair of subsets C_1 and C_2 orthogonal to another pair of subsets D_1 and D_2 (fig. 2.2) and assume that such subsets intersect in points a , b , c and d . a , b, c and d are the vertices of a so-called hyperrectangle in H .

It is easy to prove that any two opposite edges of a hyperrectangle are equal. Indeed

(2.6.7) $$c - d = (a-d) + (b-a) + (c-b)$$

and thus, as (a - d) and (c - b) are orthogonal to (c - d),

$$(c-d, c-d) = (a-d+b-a+c-b, c-d) = (b-a, c-d). \qquad (2.6.8)$$

Similarly,

$$(b-a, b-a) = (b-a, c-d) \qquad (2.6.9)$$

and thus

$$(b-a, b-a) = (c-d, c-d). \qquad (2.6.10)$$

By virtue of (2.6.8 - 10),

$$\begin{aligned}((b-a)-(c-d), (b-a)-(c-d)) &= (b-a, b-a) - \\ -2(b-a, c-d) + (c-d, c-d) &= 0\end{aligned} \qquad (2.6.11)$$

and thus

$$c-d = b-a. \qquad (2.6.12)$$

Introducing (2.6.12) in (2.6.7) there results immediately

$$c-b = d-a. \qquad (2.6.13)$$

Subsets C_1 and C_2 or D_1 and D_2, may be called parallel to each other. The constant difference between elements of two parallel subsets belonging to a subset orthogonal to both will be called the difference between the two parallel subsets and the norm of such difference the distance between both.

If two parallel subsets have a common element, their difference is zero. This implies that, if two subsets C_1 and C_2 are both orthogonal to D, and they both contain the same element x, the intersections e_1 and e_2 of C_1 and D and of C_2 and D, if they exist, cannot be distinct.

The unique intersection of a given subset C and a subset D orthogonal to C and containing a given point x is called the orthogonal projection, or simply the projection of x in C.

2.7 Local Concepts

Before proceeding further a remark must be done concerning the notation which will be used in the sequel.

Two kinds of inner-product will be associated with a Banach space X to build a Hilbert space. The first inner product will be denoted by $(\cdot,\cdot)$ and the second by $[\cdot,\cdot]$ The space built with the first will be denoted by X_H and with the second by H.
The norm connected with the first will be denoted by $\|\cdot\|$ and the norm connected with the second by $|\cdot|$. In what concerns the distances we shall have respectively $\delta(x_1,x_2)$ and $d(x_1,x_2)$ Thus,

$$\delta(x_1, x_2) = \| x_1 - x_2 \| = \sqrt{(x_1 - x_2, x_1 - x_2)} \qquad (2.7.1)$$

$$d(x_1, x_2) = | x_1 - x_2 | = \sqrt{[x_1 - x_2, x_1 - x_2]}\,. \qquad (2.7.2)$$

If a simple Banach space X is considered, and not an inner product space, then the norm will be $\| \cdot \|$ and the distance connected with such norm $\delta(\cdot, \cdot)$.

For reasons which will appear clear in Section 4.1, $[\cdot, \cdot]$ will be termed energy product [11], $|\cdot|$ the energy norm and $d(\cdot, \cdot)$ the energy distance.

Let now $f(x)$ be a function with domain X and range Y and let X and Y be Banach spaces. A norm is thus assumed to be defined in each of them. We shall denote by $\delta(x_1, x_2)$ the distance associated with such norm in the space X, i.e.

$$\delta(x_1, x_2) = \| x_1 - x_2 \|\,. \qquad (2.7.3)$$

Let x_0 and h be arbitrary elements of X and let a linear (linear in h) operator exist $df(x_0, f)$, also with domain X and range Y, such that

$$f(x_0 + h) - f(x_0) = df(x_0 + h) + \alpha(x_0, h) \qquad (2.7.4)$$

where $\dfrac{\| \alpha(x_0, h) \|}{\| h \|} \to 0$ as $\| h \| \to 0$. $df(x_0, h)$ is called the strong differential or Fréchet differential of the function $f(x)$ in the point x_0, for the increment $h \cdot \alpha(x_0, h)$ is called the remainder [22].

The strong differential does not exist sometimes and another differential called the weak differential or Gateaux differential [22] defined by

$$Df(x_0, h) = \frac{d}{dt} f(x_0 + th) = \lim_{t \to 0} \frac{f(x_0 + th) - f(x_0)}{t} \tag{2.7.5}$$

may still exist. The weak differential exists whenever the strong differential exists but the converse is not true. We observe that $Df(x_0, h)$ is not necessarily linear in h.

The second order Gateaux and Fréchet differentials can be constructed by applying (2.7.4) and (2.7.5) to the first order differential already defined.

Consider now the set C of the elements of X which satisfy the equation

$$f(x) = g \tag{2.7.6}$$

in which g is a fixed element in Y. Let x_0 belong to C and let the function $f(x)$ be differentiable in the point x_0. The set of all the elements $h \in X$ for which

$$df(x_0, h) = 0 \tag{2.7.7}$$

will be denoted by S_{x_0}. As $df(x_0,h)$ is linear in h, S_{x_0} is a linear subspace of X.

The set of all the elements x_0+h, with $h \in S_{x_0}$, i.e., the coset of S_{x_0} which contains x_0, is called the linear tangential manifold T_{x_0} of C in the point x_0 [22].

Supposing now that X is an inner product space, H, it can be proved that there exists a positive number ε such that the set of the points of C contained in the neighborhood ε of x_0 is homeomorphic to the set of their orthogonal projections in T_x, and the distance between each point in C and its orthogonal projection p_x in T_{x_0} is of higher order than the distance of x and p_x to x_0 [22].

The topological mapping of the neighborhood ε of C onto its orthogonal projection in T_{x_0} is moreover almost isometric in the point x_0, i.e., the distance of two elements x_1 and x_2 of C contained in the neighborhood ε of x_0, but otherwise arbitrary, is related to the distance of their orthogonal projections p_{x_1} and p_{x_2} in T_{x_0} by

$$d(x_1,x_2)(1-\alpha) \leq d(p_{x_1},p_{x_2}) \leq d(x_1,x_2)(1+\alpha) \qquad (2.7.8)$$

in which α is a positive number which tends to zero with $\varepsilon \cdot d(x_1,x_2)$ denotes the distance connected with the inner product by (2.5.5) [22].

Consider now two subsets C and D of a Hilbert space X_H, and let x_0 be a point common to both. As-

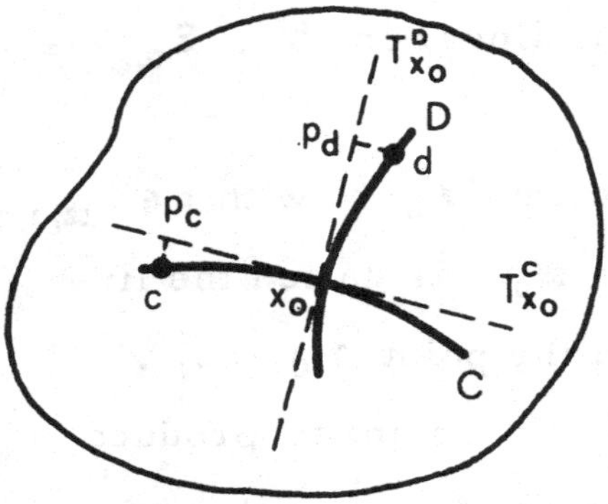

Fig. 2.3

sume that the linear tangential manifolds of C and D in the point x_0 are orthogonal. Then C and D will be called locally orthogonal (fig. 2.3).

Let c and d be points respectively of C and D . Let p_c and p_d denote their projections respectively in $T^C_{x_0}$ and $T^D_{x_0}$. Then

$$[p_d - x_0 , p_c - x_0] = 0 . \tag{2.7.9}$$

and, by virtue of Pythagoras theorem

$$d^2(x_0, p_c) + d^2(x_0, p_d) = d^2(p_c, p_d). \tag{2.7.10}$$

Considering all that has been said above, and assuming that c and d are contained within the neighborhood ε of x_0, it may be shown that

$$d^2(c,d)(1-\alpha) \leq d^2(x_0,c) + d^2(x_0,d) \leq d^2(c,d)(1+\alpha)$$

(2.7.11)

where α is a positive number which tends to zero with ε .

2.8 Extrema of Functionals

Consider the equation

$$P(x) = 0 \qquad (2.8.1)$$

where P denotes a continuous non-linear operator. Both the domain and range of the operator are assumed to be contained in a Hilbert space X_H.

The operator P is called the weak gradient (°) of the functional $F(x)$ and we write

$$P(x) = \text{grad } F(x) \qquad (2.8.2)$$

if

$$(P(x), h) = \lim_{t \to 0} \frac{F(x + th) - F(x)}{t} \qquad (2.8.3)$$

This means that the Gateaux differential $DF(x,h)$ equals the inner product of grad $F(x)$ and h :

$$DF(x,h) = (\text{grad } F(x), h) \qquad (2.8.4)$$

The operator is said to be a potential operator

(°) The strong gradient can also be defined if the function is Fréchet differentiable.

on some subset $C \subset H$ if there exists a functional $F(x)$ such that (2.8.2) holds for any $x \in C$.

A point $x_0 \in C$ is called a critical point of the functional $F(x)$ if

$$\text{grad } F(x_0) = 0. \tag{2.8.5}$$

Solving the equation (2.8.1) is thus the same as finding the critical points of functional F. Equation (2.8.1) is called the Euler equation for functional $F(x)$.

If $F(x) \geq F(x_0)$ for all $x \in C_0 \subset C$, then x_0 is a global minimizer of $F(x)$ on C_0.

If an open neighborhood N of x_0 exists such that $F(x) \geq F(x_0)$ for all x contained in $N \cap C_0$, then x_0 is a local minimizer of F. It can be shown that a local minimizer is a critical point of F.

Both a global minimizer or a local minimizer will be called proper if x_0 is the unique minimizer, i.e. if the definition implies $F(x) > F(x_0)$ and not $F(x) \geq F(x_0)$.

A critical point of $F(x)$ will be a proper local minimizer if the second differential $d^2F(x_0,h,h)$ is positive definite (i.e. $d^2F(x,h,h) > 0$ for all $h \subset N \cap C_0$) at x_0. If $d^2F(x_0, h,h)$ is positive semi-definite then the critical point will be no more than a local minimizer.

Similar definitions can be introduced formaximizers. The second differential is then negative definite or negative semi-definite.

Maximizers and minimizers are called extreme points or extrema of the functional.

It can be demonstrated that if $x \in X$, $f(x)$ is a function with domain X and range Y, $F(x)$ is a functional which has a minimum in x under the condition $f(x) = g$, where $g \in Y$ then there exists a linear functional ℓ, defined on Y, such that x_0 is also a critical point for the functional

$$J(x) = F(x) - \ell\left[f(x) - g\right]. \qquad (2.8.6)$$

This is the basis of the Lagrangian multiplier method.

3. THE CONVERGENCE THEORY FOR VARIATIONAL METHODS

3.1 Isoconstrained and Isominimizing Subsets

Let X be a Banach space and let $\mathfrak{F}$ denote a family of continuous functionals on X.

Assume that each functional of the family admits a proper global minimizer in a certain class of subsets of X, called the isoconstrained subsets of X. This means that each isoconstrained subset $C \subset X$ contains an element such that

$$F(\mathfrak{s}) < F(c) \tag{3.1.1}$$

in which c denotes any element in C distinct from $\mathfrak{s}$.

The isoconstrained subsets are assumed to be homeomorphic to a certain subspace of X and the union of all the isoconstrained subsets is assumed to coincide with the whole space X.

Let us consider now, for each functional F of the family $\mathfrak{F}$, the set of all the proper global minimizers corresponding to all the isoconstrained subsets of X. We call such set the isominimizing subset of X corresponding to F and assume the family $\mathfrak{F}$ to be such that the union of all the isominimizing subsets corresponding to all the different functionals of $\mathfrak{F}$ coincides with the whole space, X.

The intersection of each isominimising and

each isoconstrained element is assumed to contain one, and no more than one, element.

Two elements belonging to the same isoconstrained (or isominimizing) subset are called isoconstrained (or isominimizing) elements.

Consider now a bounded continuous operator T with domain X and range X', X' denoting a subspace of X. To each element in X there corresponds an element in X', although more than one element in X may correspond to each element in X'.

It is assumed that the T-imsge of any element belonging to X' coincides with the element itself, i.e. that

$$T(e) = e \quad \text{if} \quad e \in X'. \qquad (3.1.2)$$

It is also assumed that isoconstrained and isominimizing subsets can be defined also on X', with respect to the same family of functionals, $\mathfrak{F}$. Such isoconstrained and isominimizing subsets are assumed to meet the same requirements as those in X.

We assume further that the T-images of the elements belonging to any isoconstrained subset C of X are contained in an isoconstrained subset C' of X', which is said to correspond to C. This means that the T-images of any two isoconstrained elements, e and f, in X are isoconstrained elements, e' and f', in X' (fig. 3.1), although they are not necessarily isoconstrained elements in X.

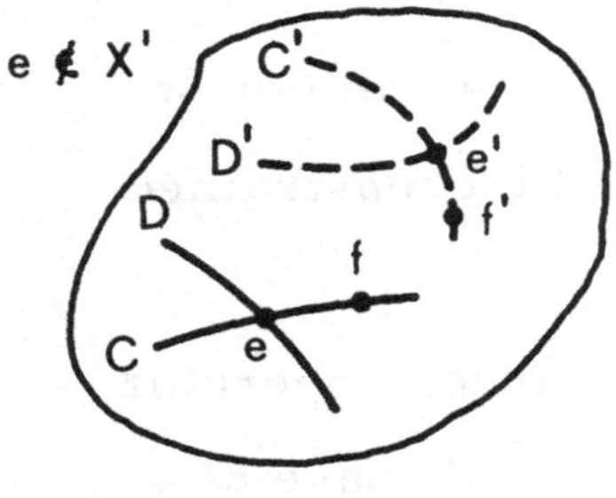

Fig. 3.1

If, however, any two isoconstrained elements in X' are isoconstrained elements in X, the operator T is said to be conforming (°).

It is clear that, if a given element e belongs to X' as well as to X, then, as, by virtue of (3.1.2), it coincides with its own T-image, it must be contained both in C and in C' (fig. 3.2).

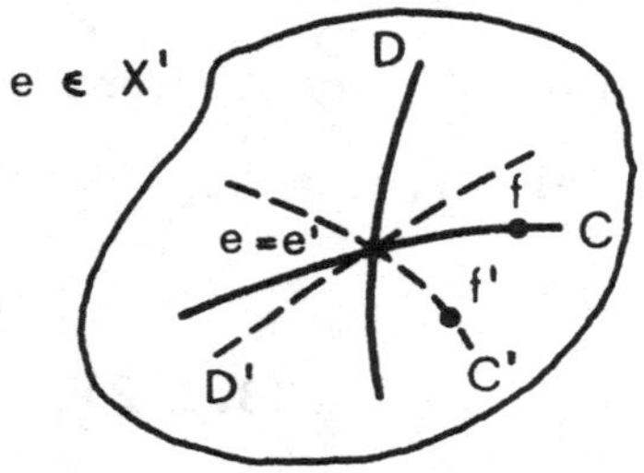

Fig. 3.2

A given isominimizing subset D' of X' is said to correspond to a certain isominimizing subset D of X if they both correspond to the same functional.

A second operator, A, can thus be considered, with domain X and range X', which makes the intersection of each isoconstrained and each isominimizing subset of X correspond to the intersection of the corresponding isoconstrained and isominimizing subsets of X'.

(°) As the zero element belongs both to X and X', $T(0) = 0$. The isoconstrained subset of X' which contains the zero element must thus be contained by the corresponding isoconstrained subset of X, if T is conforming.

A is assumed also bounded and continuous, and also such that

$$A(e) = e \quad \text{if} \quad e \in X' \tag{3.1.3}$$

This means that on each element of X' cross an isoconstrained and an isominimizing subset of X and the corresponding isoconstrained and isominimizing subsets of X' (fig. 3.2).

The A-image of an element e is called the approximation of e in X'.

We remark that the A-images of any two isominimizing elements of X are isominimizing elements of X'. Besides, the A-images of any two isoconstrained elements of X are also isoconstrained elements of X'.

3.2 Space of Constraints and Functionals

Let $\mathcal{G}$ denote the set of all the isoconstrained subsets of X. We assume $\mathcal{G}$ to be a Banach space and call it the space of constraints.

Let $g(x)$ be the function, assumed to be continuous, which associates to each element of X a given element in $\mathcal{G}$. Each isoconstrained subset G of X is thus the set of all the points $x \in X$ which verify the equation

(3.2.1) $$g(x) = G$$

where G is a fixed point in $\mathcal{G}$.

According to the assumptions introduced in the preceding Section, the T-images of the elements of G are isoconstrained elements in X'. Let G' the isoconstrained subset of X' which contains such T-images of the elements of G, i.e. the isoconstrained subset of X' corresponding to G. G may still contain the T-images of the elements of other isoconstrained subsets of X. Let $\mathcal{G}_{G'}$ denote the set of the isoconstrained subsets of X whose T-images belong to G'. $\mathcal{G}_{G'}$ is a subset of $\mathcal{G}$.

Let now $\mathcal{G}'$ denote the set of all the subsets $\mathcal{G}_{G'}$ of $\mathcal{G}$ corresponding to the different isoconstrained subsets of X'. We assume $\mathcal{G}'$ to be also a Banach space and call it the generalized space of generalized constraints.

Let $g'(x')$ be the function, assumed to be continuous, which associates to each element of X' a given element in $\mathcal{G}'$. Each isoconstrained subset G' of X' is thus the set of all the points $x' \in X'$ which verify the equation

(3.2.2) $$g'(x') = G'$$

where G' is a fixed point in $\mathcal{G}'$.

Assume now that the family of functionals $\mathcal{F}$ forms also a Banach space called the space of functionals. As each element x in X corresponds to a given isominimizing subset and such isominimizing subset corresponds to a given functional, we can consider the function $f(x)$, assumed continuous, which associates to each element x in X a given element of $\mathfrak{F}$. Each isominimizing subset F of X is thus the set of all the points $x \in X$ which verify the equation

$$f(x) = F \tag{3.2.3}$$

where F is a fixed element in $\mathfrak{F}$.

According to the assumptions introduced in 3.1, the A-images of the elements of F are isominimizing elements in X'. Let F' be the isominimizing subset of X' which contains such A-images of the elements of F.

F' may still contain the A-images of the elements of other isominimizing subsets of X. Let $\mathfrak{F}_{F'}$ denote the set of the isominimizing subsets of X whose A-images belong to F'. $\mathfrak{F}_{F'}$ is a subset of $\mathfrak{F}$.

Let now $\mathfrak{F}'$ denote the set of all the subsets $\mathfrak{F}_{F'}$ of $\mathfrak{F}$ corresponding to the different isominimizing subsets of X'. We assume $\mathfrak{F}'$ to be also a Banach space and call it the space of generalized functionals.

Let $f'(x')$ be the function, assumed continuous, which associates to each element of X' a given element in $\mathfrak{F}'$. Each isominimizing subset F' of X' is thus the set of all the points $x' \in X'$ which verify the equation

$$f'(x') = F' \tag{3.2.4}$$

where f' is a fixed point in $\mathfrak{F}'$.

Assume now that function $g(x)$ is linear. Then, the isoconstrained subsets of X are the cosets of a linear subspace of X with equation

$$g(x) = 0. \tag{3.2.5}$$

Indeed, if x_1 and x_2 are isoconstrained elements in X we have

$$g(x_1) = g(x_2) = G \tag{3.2.6-7}$$

and then

$$g(x_1 - x_2) = g(x_1) - g(x_2) = 0. \tag{3.2.8}$$

This means that $x_1 - x_2$ belongs to the isoconstrained subset of X with equation (3.2.5) which is a linear subspace of X because, as g is linear, if x_1 and x_2 verify (3.2.5), $\alpha x_1 + \beta x_2$ (α and β , real numbers) also verifies (3.2.5).

It could similarly be shown that, if g' is linear, the isoconstrained subsets of X' are the cosets of a linear subspace of X' with equation

$$g'(x') = 0 \tag{3.2.9}$$

and analogous conclusions concerning the isominimizing subsets could be drawn if functions f and f' were linear.

We remark that, if the isoconstrained subsets of X are cosets of a linear subspace of X, function $g(x)$ is not necessarily a linear function. The only conclusion which can be drawn about $g(x)$ then is that $g(x_1 - x_2) = 0$ whenever $g(x_1) = g(x_2)$. Analogous remarks can be made about $g'(x')$, $f(x)$ and $f'(x')$.

We further remark that, if the isoconstrained subsets of X are cosets of a linear subspace of X, and T is is a linear operator, then the isoconstrained subsets of X' are cosets of a linear subspace X'. On the other hand, if the isominimizing subsets of X are cosets of a linear subspace of X, and A is a linear operator, then the isominimizing subsets of X' are cosets of a linear subspace of X'.

We define now the linear tangential manifolds of the isoconstrained and isominimizing subsets in terms of the functions g, g', f and f'.

The linear tangential manifold T_x^G in a point x of the isoconstrained subset $G \subset X$ is the coset which contains x of the linear subspace S_x^G of X with equation

$$dg(x,h) = 0 \qquad (3.2.10)$$

where h is any point belonging to S_x^G.

The linear tangential manifold $T_{x'}^{G'}$ in a point x' of the isoconstrained subset $G' \subset X'$ is the coset which contains x' of the linear subspace $S_{x'}^{G'}$ of X' with equation

$$dg'(x', h') = 0 \qquad (3.2.11)$$

where h' is any point belonging to $S_{x'}^{G'}$.

The linear tangential manifold T_x^F in a point x of the isominimizing subset $F \subset X$ is the coset which contains x of the linear subspace S_x^F of X with equation

$$df(x, k) = 0 \qquad (3.2.12)$$

where k is any point belonging to S_x^F .

The linear tangential manifold $T_{x'}^{F'}$ in a point x' of the isominimizing subset $F' \subset X'$ is the coset which contains x' of the linear subspace $S_{x'}^{F'}$ of X' with equation

$$df'(x', k') = 0 \qquad (3.2.13)$$

where k' is any point belonging to $S_{x'}^{F'}$.

3.3 The Approximation Theorems

The approximation theorems present sufficient conditions for the smallness of the distance between a given element in X and its approximation in X' .

Let s denote the proper global minimizer of F on the isoconstrained subset C of X . Let s'_a denote the unique global minimizer of F on the corresponding isoconstrained subset C' of X'. s'_a is thus the approximation of s.

A Lema will be introduced first.

Lema: "Given an arbitrary positive number δ, it is possible to determine another arbitrary positive number δ' such that, if c is a point belonging to C and $F(c) - F(s) < \delta$ the distance between c and s is smaller than δ ".

Let us consider indeed a sphere S_δ in C with radius δ and center s and let F_δ be the minimum value which F takes on $C - S_\delta$ i.e. outside the sphere. By virtue of (3.1.1),

$$F_\delta - F(s) > 0 . \qquad (3.3.1)$$

Thus, given any point c such that

$$F(c) < F_\delta \qquad (3.3.2)$$

the distance $d(s, c)$ between s and c must be smaller than δ. This means that δ' can be equal to $F_\delta - F(s)$, i.e. δ' can be the difference between the minimum values of F on C, inside and outside S_δ.

A second Lema might result immediately from adapting the first one to the space X'.

The first approximation theorem states that: "Given an arbitrarily small positive number δ, it is possible to determine another arbitrarily small positive number ε such that, if

a) the operator T is conforming and $C' \subset C$,

b) one point c' exists on C' whose distance to s is smaller than ε, then the distance $d(s, s'_a)$ is smaller than δ''.

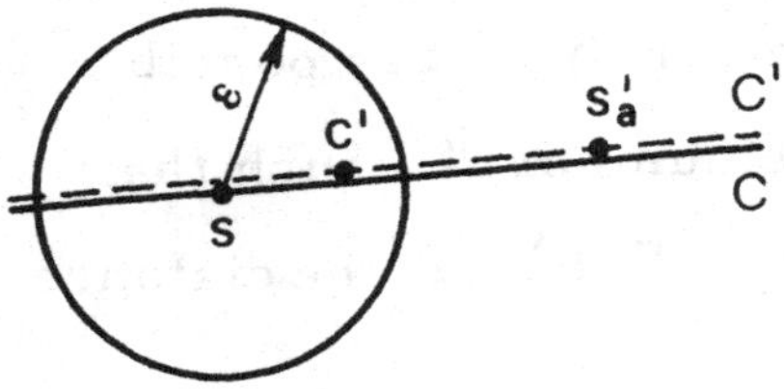

Fig. 3.3

Fig. 3.3 represents graphically the situation.

As $C' \subset C$, and $s'_a \in C'$,

$$F(s) \leq F(s'_a) \tag{3.3.3}$$

As $c' \in C'$

$$F(s'_a) \leq F(c'). \tag{3.3.4}$$

By virtue of condition b),

$$d(s, c') < \varepsilon . \tag{3.3.5}$$

Now, as F is continuous on X, it is possible, given an arbitrarily small positive number, ε', to determine ε such that (3.3.5) implies

$$|F(c') - F(s)| < \varepsilon'. \tag{3.3.6}$$

Combining (3.3.3) and (3.3.4) we obtain

$$F(s) \leq F(s'_a) \leq F(c') \tag{3.3.7}$$

and, by virtue of (3.3.6),

$$F(s) \leq F(s'_a) < F(s) + \varepsilon' \tag{3.3.8}$$

which implies

$$F(s'_a) - F(s) < \varepsilon'. \qquad (3.3.9)$$

Now, s'_a belongs to C' and $C' \subset C$ and then it is possible, by virtue of the Lema to determine ε', and then ε, in such a way that

$$d(s, s'_a) < \delta. \qquad (3.3.10)$$

Let us consider now the second approximation theorem:

"Given an arbitrarily small positive number, δ it is possible to determine two other arbitrarily small positive numbers ε and ε' such that if

a) one point c' exists on C' whose distance to s is smaller than ε,

b) one point c exists on C whose distance to s'_a is smaller than ε',

then, the distance between s and s'_a is smaller than δ".

The situation is graphically described by fig. 3.4.

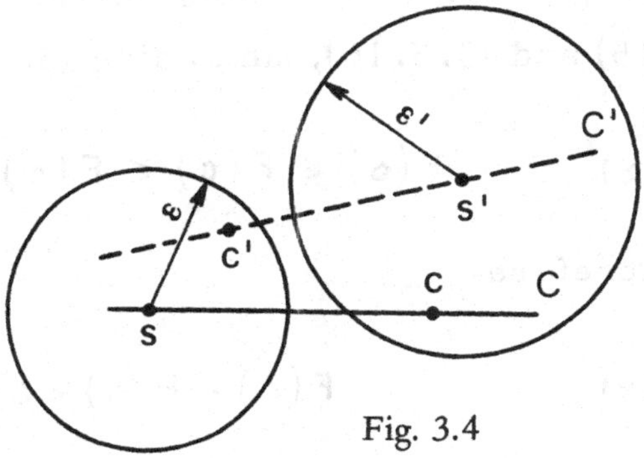

Fig. 3.4

As $c \in C$

(3.3.11) $$F(s) \leq F(c)$$

and, as

(3.3.12) $$F(s'_a) \leq F(c').$$

By virtue of conditions b) and c),

(3.3.13) $$d(s, c') < \varepsilon$$

(3.3.14) $$d(c, s'_a) < \varepsilon'$$

As F is a continuous functional on X, it is possible to choose ε and ε' in such a way that

(3.3.15) $$F(s) = F(c') \pm \varepsilon''$$

(3.3.16) $$F(c) = F(s'_a) \pm \varepsilon'''$$

(3.3.17) $$\varepsilon' < \delta$$

ε'' and ε''' being given arbitrarily small positive numbers.

Eliminating $F(s'_a)$ and $F(c')$ from (3.3.12), (3.3.15) and (3.3.16), and using (3.3.11), we obtain

(3.3.18) $$F(s) \leq F(c) \leq F(s) \pm \varepsilon''' \pm \varepsilon''$$

and therefore

(3.3.19) $$F(c) - F(s) < \varepsilon'' + \varepsilon'''$$

Now, according to the Lema, it is possible to determine $\varepsilon'' + \varepsilon'''$, and then ε and ε', in such a way that

$$d(c, s) < \delta - \varepsilon'. \tag{3.3.20}$$

The triangular inequality permits to write

$$d(s, s'_a) \leq d(s, c) + d(c, s'_a) \tag{3.3.21}$$

and thus, by virtue of (3. 3. 14) and (3. 3. 20),

$$d(s, s'_a) < \delta . \tag{3.3.22}$$

As it will be seen in the next Section, the element $c' \in C'$ which, in both theorems, is required to lie within the neighbourhood ε of s is usually the T-image of s, $T(s)$.

In what concerns the element $c \in C$ which, in the second theorem, is required to lie within the neighborhood ε' of s'_a, it is fair to say that sometimes it is not easy to show that it exists.

It may be convenient to consider as element c an element s_a of C such that $T(s_a) = s'_a$.

Another possibility arises when X can be made an inner-product space H and the isoconstrained and isominimizing subsets are orthogonal subsets of H.

3.4 Projection Theorem

Let us build a Hilbert space H by associating the inner-product $[e, f]$ to each pair of elements e and f of X . Let H' be the subspace of H corresponding to X' .

Assume that the isoconstrained and isomonimizing subsets of H are orthogonal or, more precisely, that they are cosets of two orthogonal subspaces of H . Assume the same about the isoconstrained and isominimizing subsets of H' (°). The following theorem holds:

"Let p be the projection of s'_a in C and p' the T -image of p in H' . Let $\|A\|$ be the norm of the operator A . Then

(3.4.1) $$d(s'_a, p) < (1 + \|A\|)\, d(p, p')"$$

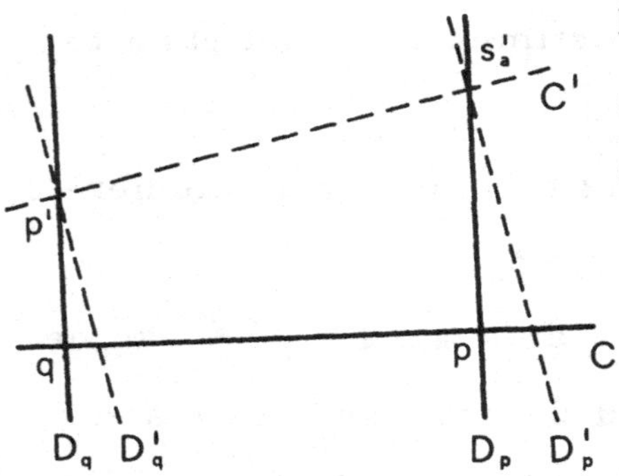

Fig. 3.5

Indeed, as shown by fig. 3.5, p' and s'_a are both on C'. p and q are projections of s'_a and p' on C .

(°) This will happen if f and g introduced in Section 3.2 are assumed to be linear as well as operator T .

By Pythagora's theorem, (°)

$$d^2(p,q) + d^2(p',q) = d^2(p,p') \tag{3.4.2}$$

Then

$$d(p,q) \leq d(p,p'). \tag{3.4.3}$$

Now, as D_q and D'_q cross on p', and p' belongs to H', D'_q is the isominimizing set of H' which corresponds to M_q. The same can be said about D_p and D'_p because they cross on s'_a. And, as p' and s'_a are contained in C', which is the isoconstrained subset of H' which corresponds to C, there follows that p' and s'_a are the approximations of q and p. Hence

$$d(p', s'_a) < \| A \| \, d(p,q) \tag{3.4.4}$$

or, by virtue of (3.4.3),

$$d(p', s'_a) < \| A \| \, d(p,p'). \tag{3.4.5}$$

But by the triangular inequality,

$$d(s'_a, p) \leq d(s'_a, p') + d(p,p') \tag{3.4.6}$$

(°) The distance between two points e and f in H is of course assumed to equal $\sqrt{[e-f, e-f]}$.

and thus, by virtue of (3.4.5), we obtain (3.4.1).

Using this theorem, condition b) of the second approximation theorem can be modified. It can be replaced indeed by the requirement that the distance between the projection p of s'_a on C and its T-image p' in H' be smaller than an arbitrarily small positive number.

3.5 Generalization of the Projection Theorem

In the general case, the isoconstrained and isominimizing subsets are not cosets of linear subspaces of the space X, and this means that no inner-product can be introduced which makes them orthogonal.

However, very often, it is still possible to define the inner-product in such a way that, given a point $x \in X$, the linear tangential manifolds of the isoconstrained and isominimizing manifolds which cross at x are orthogonal.

Let such inner-product be denoted by $[\ ,\ ]_x$. Let H_x denote the Hilbert space which is obtained by associating this definition to the elements of X. Let $d_x(\cdot,\cdot)$ be the distance connected to the inner-product $[\ ,\ ]_x$.

By virtue of (2.7.9) we can write for two points c and d contained within the neighborhood ε of x and belonging respectively to the isoconstrained and isominimizing subsets

sets which cross at x ,

$$d_x^2(c,d)(1-\alpha) \le d_x^2(x,c)+d_x^2(x,d) \le d_x^2(c,d)(1+\alpha) \tag{3.5.1}$$

where α is a positive number which tends to zero with ε . Let us define a metric in the space X by requiring first

$$d_x(a,x)(1-\alpha) \le d_x(a,x) \le d_x(a,x)(1+\alpha) \tag{3.5.2}$$

where α tends to zero with $d_x(a,x)$, and then defining the distance between two points as the smallest obtained by adding the infinitesimal distances along all possible continuous pathes between x_1 and x_2 , i.e.,

$$d(x_1,x_2) = \min \int_{x_1}^{x_2} ds \tag{3.5.3}$$

in which ds denotes the distance (defined through (3.5.2) between two points very near each other on a given path.

We observe that we can speak about continuous pathes because a definition of distance has already been introduced before. Such definition can be any of the distances $d(\cdot,\cdot)_x$ or then the distance $\delta(\cdot,\cdot)$ equal to the norm $\|\cdot\|$.

Considering (3.5.2) and (3.5.1) it becomes clear that for every point x of X , and for any two points c and d in the neighborhood ε of x , we can write

$$d^2(c,d)(1-\beta) \le d^2(x,c)+d^2(x,d) \le d^2(c,d)(1+\beta) \tag{3.5.4}$$

where β tends to zero with ε .

Using this result, the projection theorem demonstrated in the preceeding Section can easily be generalized. The generalized statement will coincide with the one already presented however, if the expression "projection p of s'_a in C " is assumed to mean "the intersection with C of the iso-minimizing subset which contains s'_a ". Such meaning will be adopted in the sequel.

3.6 Convergence Theorems

Sequences of continuous bounded operators $\{T_n\}$ and $\{A_n\}$ will be considered now. To each term of these sequences a subspace $X'_n \subset X$ will correspond which is the range of the operators T_n and A_n .

A sequence of subsets C'_n of the subspaces X'_n is called complete with respect to another subset S of C if, given a positive arbitrarily small number , ε, it is possible to determine a positive integer N such that for any $n > N$ at least one element of C'_n exists within the neighborhood ε of each element belonging to S [11, 13, 17] .

The convergence theorems can be based in such a general concept of completeness. The discussion of convergence will be easier however if the element in C'_n which approximates a given element of S is its T_n -image in X'.

The concept of completeness which will be implicitly used throughout this paper is thus a particularization of the general one which may be called T -completeness. A sequence of subsets of $C'_n \subset C$ will be called T -complete with respect to another subset S of C if, given ε it is possible to determine N such that for $n > N$, the distance between any element e in S and its T_n -image in X'_n is smaller than ε, i.e., if

$$d\left(e, T_n(e)\right) < \varepsilon \tag{3.6.1}$$

We observe that if the sequence $\{C'_n\}$ is T-complete, it will be just complete, although the converse is not necessarily true.

The convergence theorems can be stated for different degrees of generality. For each degree of generality a pair of theorems occurs each of them is related to the two approximation theorems of the preceeding Section.

Two different degrees of generality will be considered in the present paper corresponding to two different situations.

In what is called the first situation, a unique space X, a unique family $\mathfrak{F}$ and a unique element $e \in X$ will be considered.

The first approximation theorem gives then

rise to the following convergence theorem: "If

a) the operators T_n are conforming and $C'_n \subset C$ (conformity condition),

b) the sequence $\{C'_n\}$ is complete with respect to a subset S of C which contains $\mathfrak{s}$ (completeness condition)

then $\{\mathfrak{s}'_{an}\}$ converges to $\mathfrak{s}$ ".

The conditions a) and b) correspond to conditions a) and b) of the first approximation theorem and no comment is needed.

The second convergence theorem, which results from the second approcimation theorem, is the following one: "If

a) the sequence $\{C'_n\}$ is complete with respect to a subset S of C which contains $\mathfrak{s}$ (completeness condition),

b) $\{\mathfrak{s}'_{an}\}$ converges to an element of C,

then, $\{\mathfrak{s}'_{an}\}$ converges to $\mathfrak{s}$ ".

According to what was said in the preceding Section, condition b) will be verified if a sequence of elements $\mathfrak{s}_{an}$ belonging to C, such that $T_n(\mathfrak{s}_{an}) = \mathfrak{s}'_{an}$, converges to an element of S. Or, in the local orthogonal case, if the sequence of the projections (see Section 3.4) of the elements $\mathfrak{s}'_{an}$ on C converges to an element of S.

In what is called the second situation, not a unique space X, a unique family $\mathfrak{F}$ and a unique element $e \in X$ but a complete metric linear space $\mathfrak{X}$ and a sequence of subspaces $X_n \subset \mathfrak{X}$, a sequence of families $\mathfrak{F}_n$ and a sequence of elements $e_n \in X_n$ are to be considered. The domain of each operator T_n or A_n is X_n. The range is X'_n as before.

It is convenient to generalize now, the concept of completeness. A sequence of subsets C'_n of the subspaces X'_n will be called complete with respect to a sequence of subsets S_n of the subspaces C_n if, given a positive arbitrarily small number ε, it is possible to determine a positive integer N such that for any $n > N$, at least one element of C'_n exists within the neighborhood ε of each element of S_n.

The first convergence theorem becomes now:

"If a) the operators T_n are conforming and $C'_n \subset C_n$,

b) the sequence $\{C'_n\}$ is complete with respect to a sequence of subsets S_n of the subspaces C_n such that $e_n \in S_n$,

then, $\{s'_{an}\}$ converges to the same limit as $\{s_n\}$",

and the second convergence theorem becomes:

"If a) the sequence $\{C'_n\}$ is complete with respect to a sequence of subsets S_n of the subspaces C_n such that $e_n \in S_n$,

b) there exists a sequence of elements of the different spaces C_n which tends to the same limit as $\{s'_{an}\}$

then $\{s'_{an}\}$ converges to the same limit as $\{s_n\}$".

Condition b) will be verified if a sequence of elements of the different subsets exists which has the same limit as a sequence of elements s belonging to the different subspaces C_n, such that $T_n(s_{an}) = s'_{an}$ or, in the orthogonal case, the same limit as the sequence of the projections p_n of each element s'_{an} on the different subspaces C_n .

3.7 Discretization Techniques

The conversion of a given variational problem into a new one, eventually simpler than the first and capable of providing an approximate solution to it has been discussed along this chapter.

If operator T is chosen in such a way that X' is N-dimensional, then the variational problem becomes particularly easy to solve as the functional F becomes a function of N variables and the element which minimizes the functional can be determined by just solving the system of N equations to N unknowns which results from equating to zero the derivatives of the function with respect to each variable. The problem is said to have been discretized. Of course the equations will be linear only if the functional F is a quadratic function of the variables.

We observe that completeness can be achieved only if X is a separate space.

If the operator T is conforming and $C' \subset C$, this discretization procedure is nothing else than the classical Ritz method [11] to which the first convergence theorem can be applied.

If the operator T is not conforming, then we

must resort to the second theorem. Such is very often the case of the finite element method which will be discussed in the sequel.

3.8 Variational Theorems

Suppose that the proper global minimizer of a given functional of the family $\mathfrak{F}$ on a given isoconstrained subset C of X is also a proper local minimized and thus a critical point of the functional.

The restriction to C can be avoided by using Lagrangian multipliers according to Section 2.8. The same point s of X which minimized F on C is indeed a critical point of the functional

$$J(x) = F(x) - \ell\{g(x) - G\} \tag{3.8.1}$$

on the whole space X, supposing that the isoconstrained subset C is characterized by equation (3.2.1).

The general expression (3.8.1) can now be particularized for other subsets of X. A system of variational theorems can thus be established, each of them referring to a given class of subsets of X and to a corresponding family of functionals which result from particularizing (3.8.1) to such class of subsets. Each variational theorem states that the critical point of a given functional on a given subset if the proper local minimizer of a certain functional of family $\mathfrak{F}$ on a certain

isoconstrained subset.
An analogous system of variational theorems can be established for space X'.

This means that the approximation of a given point $\mathfrak{s}$ of X which is a critical point for a given functional on a given subset S of X can be determined by determining the critical point of the same functional on the T-image of S in X'.

The convergence of sequences of such approximations can be examined only if such critical points are also minimizers of a certain functional on isoconstrained subsets, because then all the theory can be applied. No conclusions can be drawn however if no minimum principles are known.

4. MATHEMATICAL THEORY OF STRUCTURES

4.1 Introduction

Chapters 4, 5, 6 and 7 will be devoted to the Mathematical Theory of Structures. The present Chapter is concerned with the general lines of such theory.

There is no reason why the philosophy of the generation of the different models of the Theory of Structures should be different from the philosophy adopted in analogous situations of Mathematical Analysis.

An approximate method of Mathematical Analysis has not to be derived but described and justified. And the most general justification criterion used in Mathematical Analysis is the convergence criterion, according to which an approximate method has to be rejected if it cannot provide a sequence of approximate solutions which converges to the exact one. This is why the Mathematical Theory of Structures should consist essentially of three parts: a general model, a rule or rules to generate approximate models from an initial one, a justification of such rules.

Variational theorems have been used in the past to unify the different continuous and discrete models or theories applied in structural analysis. It was recognized however that a serious limitation of variational theorems is that they are unable to supply information on convergence, unless they are extremum principles.

The fundamental role has to be trasnferred to the convergence theorems of the general theory of variational methods.

4.2 The General Model

The general analytical model of the Theory of Structures involves three kinds of magnitudes - stresses, strains and displacements - the vectors of which will be denoted by $\underline{s}$, $\underline{e}$ and $\underline{u}$. Such magnitudes are defined on a domain Δ corresponding to the structure and interrelated by three kinds of equations which can be symbolized as follows:

(4.2.1) - equilibrium equations: $\underline{E}\,\underline{s} = \underline{f}$

(4.2.2) - strain-displacement equations: $\underline{e} = \underline{e}^{0} + \underline{D}\,\underline{u}$

(4.2.3) - stress-strain equations: $s = \dfrac{\partial W}{\partial \underline{e}}$.

$\underline{E}$ and $\underline{D}$ are linear differential operators, $\underline{f}$ is the vector of the body force density components, $\underline{e}^{0}$ is the vector of the initial strains and W (strain energy density) is a positive definite (stability assumption) function of the strains.

The linearity of operators $\underline{E}$ and $\underline{D}$ is one of the simplifying assumptions which characterize geometrical linearity. Later in this Chapter (Section 4.11) it will be shown that geometrically non-linear problems can be reduced to geometrically linear ones. The admission of such assumptions is thus not a really serious restriction.

On the boundary, the equilibrium equations become

$$\underline{N}\,\underline{s} = \underline{p} \tag{4.2.4}$$

where $\underline{N}$ is a matrix whose elements are odd linear functions of the components of the unit normal vector at each boundary point, so that their sign is reversed if thenormal vector is also reversed. $\underline{p}$ is the vector of the tractions applied to the boundary.

The initial strains $\underline{e}^{o}$ correspond to the initial stresses

$$\underline{s}^{o} = \left(\frac{\partial W}{\partial \underline{e}}\right)_{\underline{e}=\underline{e}^{o}} \tag{4.2.5}$$

i.e. to the stresses which appear in the structure when the displacements are assumed to vanish. As the displacements are assumed to vanish when no external forces are acting on the structure, $\underline{s}^{o}$ respects the equilibrium equations

$$\underline{E}\,\underline{s}^{o} = \underline{0} \tag{4.2.6}$$

$$\underline{N}\,\underline{s}^{o} = \underline{0}\,. \tag{4.2.7}$$

Given $\underline{u}$ and $\underline{e}^{o}$, it is easy to determine a unique strain vector $\underline{e}$ by using (4.2.7). On the other hand, we assume that, given $\underline{e}$, it is possible to determine a unique vector $\underline{e}^{o}$, respecting (4.2.6) and (4.2.7), and displacements $\underline{u}$ such that (4.2.2) is fulfilled. The displacements are not unique however, as to each strain vector $\underline{e}$ several displacements may correspond which differ by a rigid body motion of the structure.

In the current literature the state of the structure free from external forces is usually taken as the undeformed state of the structure. If this is done, and initial stresses are present, the stress-strain equations cease to be homogeneous, while the strain-displacement equations are always homogeneous.

A different practice is followed in the present notes which will appear to be more convenient in the sequel. Stress-strain equations (4.2.3) were thus made homogeneous by assuming the strains to vanish simultaneously with the stresses. As a consequence, the strain-displacement equations (4.2.2) become inhomogeneous.

Structural analysis reduces to finding the solution of the system of equations (4.2.1), (4.2.2) and (4.2.3) which satisfies certain boundary conditions. The simplest and most important boundary conditions can be directly expressed in terms of displacements or tractions applied to the boundary.

Let B_1 and B_2 denote the portions of the boundary where tractions and displacements are respectively prescribed. The boundary conditions will be analytically expressed by

$$\underline{N}\,\underline{\sigma} = \underline{\bar{p}} \quad \text{on} \quad B_1 \tag{4.2.8}$$

$$\underline{u} = \underline{\bar{u}} \quad \text{on} \quad B_2 \tag{4.2.9}$$

in which the upper dashes denote prescribed magnitudes.

It has been implicitly assumed so far that the stress and displacement derivatives involved in operators $\underline{E}$ and $\underline{D}$ exist everywhere in Δ .

The problem can however be formulated in more general terms if displacement discontinuities are prescribed on an internal boundary Γ , and external forces are assumed to be distributed on such boundary .

In the sequel it will be assumed that the domain Δ is decomposed into a finite number of subdomains Δ^e , with boundaries B^e . The internal boundary is defined by

$$\Gamma = \bigcup_e \left(B^e\right) - B \tag{4.2.10}$$

i.e., as the set of all the points belonging to boundaries B^e which do not belong to boundary B . At every point on the internal boundary, indices + and - are supposed to be ascribed to each side of the boundary and the normal unit vector at each point of Γ is assumed to be directed from + to -.

In the sequel, vectors $\underline{s}$ and $\underline{e}$ will be assumed always connected by the stress-strain equations (4.2.3). A pair of a strain and a stress fields connected in this way represent a structural field.

A structural field is termed compatible with respect to a given set of incompatibilities, i.e. to given initial strains $\bar{\underline{e}}^o$, displacements $\bar{\underline{u}}$ prescribed on B_2, and displacement discontinuities $\bar{\underline{g}}$ prescribed on Γ, if equation (4.2.2) is respected on each subdomain Δ^e, equation (4.2.9) on B_2, and equation

$$\underline{u}^+ - \underline{u}^- = \bar{\underline{g}} \tag{4.2.11}$$

on Γ.

A structural field will be termed equilibrated withrespect to a given set of external forces distributed on the subdomain Δ^e, on B_1 and on Γ, with densities $\bar{\underline{F}}$, $\bar{\underline{p}}$ and $\bar{\underline{h}}$, if equation (4.2.1) is respected on each subdomain Δ^e, equation (4.2.8) on B_1 and equation

$$(\underline{N}\,\underline{s})^+ + (\underline{N}\,\underline{s})^- = \bar{\underline{h}} \tag{4.2.12}$$

on Γ.

The exact solution with respect to a given set of initial strains, body forces, displacement discontinuities and external forces distributed on Γ, and boundary conditions is a structural field which is simultaneously compatible and equilibrated with respect to such magnitudes.

Assume now that $\underline{s}$ and $\underline{u}$ are such that the derivatives involved in $\underline{E}$ and $\underline{D}$ exist everywhere in Δ^e. Let $\underline{s}$ be equilibrated by external forces $\underline{f}$ and $\underline{p}$, respectively defined on Δ^e and B^e. Let $\underline{e}$ be the strains which the initial strains $\underline{e}^o$ compatibilize with $\underline{u}$. Then operators $\underline{E}$, $\underline{D}$ and $\underline{N}$ are supposed such that the following inequality holds:

$$\int_{\Delta^e} \left[\underline{s}^T (\underline{D}\, \underline{u}) - (\underline{E}\, \underline{s})^T \underline{u} \right] d\Delta = \int_{B^e} (\underline{N}\, \underline{s})^T \underline{u}\, dB . \qquad (4.2.13)$$

Such identity expresses the work principle. It can be given the alternative form

$$\int_{\Delta^e} \underline{s}^T \underline{e}\, d\Delta = \int_{\Delta^e} \underline{s}^T \underline{e}^o\, d\Delta + \int_{\Delta^e} \underline{f}^T \underline{u}\, d\Delta + \int_{B^e} \underline{p}^T \underline{u}\, dB \qquad (4.2.13')$$

where $\underline{f}$ and $\underline{p}$ denote the external forces equilibrated by $\underline{s}$ (see equations (4.2.1) and (4.2.4)) and vector $\underline{e}^o$ compatibilizes $\underline{e}$ and $\underline{u}$ though (4.2.2). In any particular structural model, the work principle can be deduced from equations (4.2.1), (4.2.2) and (4.2.4), i.e. from the strain-displacement and equilibrium equations, and the equilibrium equations from the work principle and the strain-displacement equations.

The strain energy associated with a subdomain Δ^e is

$$U = \int_{\Delta^e} W\, d\Delta . \qquad (4.2.14)$$

The complementary energy is

(4.2.15) $$U^* = \quad W^* d\Delta$$

in which the function

(4.2.16) $$W^* = \underline{\sigma}^T \underline{e} - W$$

is the complementary energy density.

Differentiation of (4.2.16) with respect to the strains yields

(4.2.17) $$\frac{\partial W^*}{\partial \underline{e}} = \underline{0}$$

which means that W^* is an exclusive function of the stresses.

Differentiating (4.2.16) with respect to the stresses there results

(4.2.18) $$\frac{\partial W^*}{\partial \underline{\sigma}} = \underline{e}\,.$$

Linearity will be achieved only if W is a quadratic function of the strains, i.e. if

(4.2.19) $$W = \frac{1}{2}\,\underline{e}^T \underline{H}\,\underline{e}$$

in which $\underline{H}$ is a symmetric positive definite (stability assumption) matrix.

The stress equations become then

(4.2.20) $$\underline{\sigma} = \underline{H}\,\underline{e}$$

and the complementary energy density

$$W^* = \frac{1}{2} \underline{\sigma}^T \underline{H}^{-1} \underline{\sigma} . \qquad (4.2.21)$$

Introducing (4.2.20) in (4.2.21), there results

$$W = W^* \qquad (4.2.22)$$

and thus

$$U = U^* . \qquad (4.2.23)$$

We observe that the general model can be formally simplified if the equations are combined between themselves in order to eliminate some kinds of unknowns.

If the stresses and strains are eliminated, then the problem appears reformulated in terms of the displacements, i.e. the unknowns become the only unknowns, and we say that the displacement method is being applied.

The elimination of stresses and strains is easy to obtain, mainly if the stress-strain equations are linear. Introducing indeed (4.2.2) in (4.2.20), and then (4.2.20) in (4.2.1), we obtain

$$\underline{E}\,\underline{H}\,\underline{D}\,\underline{u} = \underline{f} - \underline{E}\,\underline{H}\,\underline{e}^0 . \qquad (4.2.24)$$

By virtue of (4.2.6) we obtain Navier's equations

$$\underline{E}\,\underline{H}\,\underline{D}\,\underline{u} = \underline{f} . \qquad (4.2.25)$$

The boundary conditions (4.2.8) can also be easily written in terms of the displacements. Indeed, introducing (4.2.2) in (4.2.20), and then (4.2.20) in (4.2.4), we obtain

$$\underline{N}\ \underline{H}\ \underline{D}\underline{u} = \bar{\underline{p}} - \underline{N}\ \underline{H}\ \underline{e}^{0} \tag{4.2.26}$$

and, by virtue of (4.2.7),

$$\underline{N}\ \underline{H}\ \underline{D}\ \underline{u} = \bar{\underline{p}}\,. \tag{4.2.27}$$

If the displacements and strains are eliminated, the unknowns reduce to the stresses and then the stress method is said to be applied.

4.3 The Minimum Theorems

The analytical model which has been described admits two minimum theorems which will be proved simultaneously in the sequel.

Body forces and initial strains in Δ^e, boundary conditions in B_1, and B_2, displacement discontinuities and external forces distributed on Γ, are assumed to be prescribed.

The concepts of exact solution, compatibility and equilibrium are supposed to refer to such prescribed magnitudes, along this Section and in the following ones, unless the contrary is explicitly indicated.

We call total potential (complementary) energy the functional

$$T = \sum_e \left(U^e - \int_{\Delta^e} \underline{\bar{f}}^T \underline{u}\, d\Delta\right) - \int_{B_1} \underline{\bar{p}}^T \underline{u}\, dB - \int_{\Gamma} \frac{1}{2} \underline{\bar{h}}^T (\underline{u}^+ + \underline{u}^-)\, d\Gamma \qquad (4.3.1)$$

$$\left(T^* = \sum_e \left(U^{*e} - \int_{\Delta^e} \underline{s}^T \underline{\bar{e}}^0 d\Delta\right) - \int_{B_2} \underline{p}^T \underline{\bar{u}}\, dB - \int_{\Gamma} \frac{1}{2} (\underline{p}^+ - \underline{p}^-)^T \underline{\bar{g}}\, d\Gamma\right)$$

The minimum total potential (complementary) energy theorem states that the exact solution minimizes the total potential complementary energy on the class of the compatible (equilibrated) fields. Conversely, any field which minimizes the total potential (complementary) energy on the class of the compatible (equilibrated) fields is an exact solution to the equations of the model.

Let us first prove that the exact solution makes the total potential (complementary) energy stationary. Indeed,

$$\delta T = \sum_e \left(\delta U^e - \int_{\Delta^e} \underline{\bar{f}}^T \delta \underline{u}\, d\Delta\right) - \int_{B_1} \underline{\bar{p}}^T \delta \underline{u}\, dB - \int_{\Gamma} \frac{1}{2} \underline{\bar{h}}^T (\delta \underline{u}^+ + \delta \underline{u}^-)\, d\Gamma \qquad (4.3.2)$$

$$\left(\delta T^* = \sum_e \left(\delta U^{*e} - \int_{\Delta^e} \delta \underline{s}^T \underline{\bar{e}}^0 d\Delta\right) - \int_{B_2} \delta \underline{p}^T \underline{\bar{u}}\, dB - \int_{\Gamma} \frac{1}{2} (\delta \underline{p}^+ - \delta \underline{p}^-)^T \underline{\bar{g}}\, d\Gamma\right)$$

and

$$\delta U^e = \int_{\Delta^e} \left(\frac{\partial W}{\partial \underline{e}}\right)^T \delta \underline{e}\, d\Delta = \int_{\Delta^e} \underline{s}^T \delta \underline{e}\, d\Delta \qquad (4.3.3)$$

$$\left(\delta U^{*e} = \int_{\Delta^e} \left(\frac{\partial W^*}{\partial \underline{s}}\right)^T \delta \underline{s}\, d\Delta = \int_{\Delta^e} e^T \delta \underline{s}\, d\Delta\right)$$

As the variations are taken on a set of isocompatible (isoequilibrated) fields,

(4.3.4) $$\delta \underline{e}^{0} = \underline{0} \ \text{ in } \Delta^{e} \qquad \left(\underline{E}\,\delta\underline{\sigma} = \underline{0} \ \text{ in } \Delta^{e}\right)$$

(4.3.5) $$\delta \underline{u} = \underline{0} \ \text{ on } B_{2} \qquad \left(\underline{N}\,\delta\underline{\sigma} = \underline{0} \ \text{ on } B_{1}\right)$$

(4.3.6) $$\delta \underline{u}^{+} = \delta \underline{u}^{-} \ \text{ on } \Gamma \qquad \left((\underline{N}\,\delta\underline{\sigma})^{+} + (\underline{N}\,\delta\underline{\sigma})^{-} = \underline{0} \ \text{ on } \Gamma\right)$$

By virtue of (4.3.4) and (4.2.13),

$$\int_{\Delta^{e}} \underline{\sigma}^{T}\delta\underline{e}\, d\Delta = \int_{\Delta^{e}} \underline{\sigma}^{T}(\underline{D}\,\delta\underline{u})\, d\Delta = \int_{\Delta^{e}} (\underline{E}\,\underline{\sigma})^{T}\delta\underline{u}\, d\Delta + \int_{B^{e}} (\underline{N}\,\underline{\sigma})^{T}\delta\underline{u}\, dB$$

(4.3.7)

$$\left(\int_{\Delta^{e}} \delta\underline{\sigma}^{T}(\underline{D}\,\underline{u})\, d\Delta - \int_{B^{e}} (\underline{N}\,\delta\underline{\sigma})^{T}\underline{u}\, dB = \int_{\Delta^{e}} (\underline{E}\,\delta\underline{\sigma})^{T}\underline{u}\, d\Delta = \underline{0}\right).$$

On the other hand, introducing (4.3.5) and (4.3.6), there results

$$\sum_{e}\int_{B^{e}} (\underline{N}\,\underline{\sigma})^{T}\delta\underline{u}\, dB = \int_{B} (\underline{N}\,\underline{\sigma})^{T}\delta\underline{u}\, dB + \int_{\Gamma} \left[(\underline{N}\,\underline{\sigma})^{+T}\delta\underline{u}^{+} + (\underline{N}\,\underline{\sigma})^{-T}\delta\underline{u}^{-}\right] =$$

(4.3.8) $$= \int_{B_{1}} (\underline{N}\,\underline{\sigma})^{T}\delta\underline{u}\, d\Sigma + \int_{\Gamma} \left[(\underline{N}\,\underline{\sigma})^{+} + (\underline{N}\,\underline{\sigma})^{-}\right]^{T}\delta\underline{u}^{+}\, d\Gamma$$

$$\Big(\sum_{e}\int_{B^{e}} (\underline{N}\,\delta\underline{\sigma})^{T}\underline{u}\, dB = \int_{B} (\underline{N}\,\delta\underline{\sigma})^{T}\underline{u}\, dB + \int_{\Gamma} \left[(\underline{N}\,\delta\underline{\sigma})^{+T}u^{+} + (\underline{N}\,\delta\underline{\sigma}^{-})^{T}\underline{u}^{-}\right] d\Gamma =$$

$$= \int_{B_{2}} (\underline{N}\,\delta\underline{\sigma})^{T}\underline{u}\, dB + \int_{\Gamma} (\underline{N}\,\delta\underline{\sigma})^{+T}(\underline{u}^{+}-\underline{u}^{-})\, d\Gamma = \int_{B_{2}} \delta\underline{p}^{T}\underline{u}\, dB + \int_{\Gamma} \delta\underline{p}^{+T}(\underline{u}^{+}-\underline{u}^{-})\, d\Gamma.\Big)$$

Introducing (4.3.8) in (4.3.7) and then combining with (4.3.2), we obtain

$$\delta T = \sum_{e}\int_{\Delta^{e}} (\underline{E}\,\underline{\sigma} - \underline{\bar{f}})^{T}\delta\underline{u}\, \delta\Delta + \int_{B_{1}} (\underline{N}\,\underline{\sigma} - \underline{\bar{p}})\delta\underline{u}\, dB + \int_{\Gamma} \left[(\underline{N}\,\underline{\sigma})^{+} - (\underline{N}\,\underline{\sigma})^{-}\right]^{T}\delta\underline{u}^{+}\delta\Gamma$$

(4.3.9a)

$$\left(\delta T^* = \sum_e \int_{\Delta^e} \delta \underline{\sigma}^T (\underline{e} - \underline{e}^0 - \underline{D}\underline{u})\, d\Delta + \int_{B_2} \delta \underline{p}^T (\underline{u} - \underline{\bar{u}})\, dB + \int_{\Gamma} \delta p^{+T} (\underline{u}^+ - \underline{u} - \underline{\bar{g}})\, d\Gamma .\right)$$

(4.3.9b)

Considering equation (4.3.9), it becomes clear that the exact solution makes the total potential (complementary) energy stationary on the set of the compatible (equilibrated) fields.

It is also clear, considering (4.3.9a) and the arbitrariness of the displacement variation $\delta \underline{u}$ that the exact solution is the only field which makes T stationary.

An analogous conclusion concerning the complementary energy cannot however be drawn from (4.3.9b) because the stress variation $\delta \underline{\sigma}$ must satisfy the conditions (4.3.4–6). This difficulty could be removed however by resorting to the Lagrangian multiplier method in order to eliminate the constraints (4.2.1), (4.2.8) and (4.2.12).

Let us determine the second variation now. Starting from (4.3.9), we obtain

$$\delta^2 T = \sum_e \int_{\Delta^e} (\underline{E}\,\delta\underline{\sigma})^T \delta u \; \delta\Delta + \int_{B_1} (\underline{N}\,\delta\underline{\sigma})^T \delta \underline{u}\, dB + \int_{\Gamma} \left[(\underline{N}\,\delta\underline{\sigma})^+ - (\underline{N}\,\delta\underline{\sigma})^- \right]^T \delta \underline{u}\, d\Gamma +$$
$$+ \sum_e \int_{\Delta^e} (\underline{E}\,\underline{\sigma} - \underline{\bar{f}})^T \delta^2 \underline{u}\, d\Delta + \int_{B_1} (\underline{N}\,\underline{\sigma} - \underline{\bar{p}})^T \delta^2 \underline{u}\, dB + \int_{\Gamma} \left[(\underline{N}\,\underline{\sigma})^+ - (\underline{N}\,\underline{\sigma})^- \right]^T \delta^2 \underline{u}\, \delta\Gamma$$

(4.3.10)

$$\left(\delta^2 T^* = \sum_e \int_{\Delta^e} \delta \underline{\sigma}^T (\delta \underline{e} - \underline{D}\,\delta \underline{u})\, d\Delta - \int_{B_2} \delta \underline{p}^T \delta \underline{u}\, dB + \int_{\Gamma} \delta \underline{p}^+ (\delta \underline{u}^+ - \delta \underline{u}^-)\, d\Gamma + \right.$$
$$\left. + \sum_e \int_{\Delta^e} \delta^2 \underline{\sigma}^T (\underline{e} - \underline{e}^0 - \underline{D}\underline{u})\, d\Delta - \int_{B_2} \delta^2 \underline{p}^T (\underline{u}^+ - \underline{u}^-)\, dB + \int_{\Gamma} \delta^2 \underline{p}^T (\underline{u}^+ - \underline{u}^- - \underline{\bar{g}})\, d\Gamma .\right)$$

At the stationary point, the three last terms in (4.3.10) vanish and thus the second variation becomes

$$\delta^2 T^*(\underline{s}) = \sum_e \int_{\Delta^e} \delta\underline{s}^T(\delta\underline{e} - \underline{D}\,\delta\underline{u})\,d\Delta + \int_{B_2} \delta\underline{p}^T \delta\underline{u}\,dB + \int_{\Gamma} \delta\underline{p}^+ (\delta\underline{u}^+ - \delta\underline{u}^-)\,d\Gamma \Big)$$

$$\Big(\delta^2 T(\underline{s}) = \sum_e \int_{\Delta^e} (\underline{E}\,\delta\underline{s})^T \delta\underline{u}\,d\Delta + \int_{B_1} (\underline{N}\,\delta\underline{s})^T \delta\underline{u}\,dB + \int_{\Gamma} \left[(\underline{N}\,\delta\underline{s})^+ + (\underline{N}\,\delta\underline{s})^-\right]^T \delta\underline{u}\,d\Gamma$$

which by analogy with (4.3.8) may be transformed into

$$\delta^2 T(\underline{s}) = \sum_e \left[\int_{\Delta^e} (\underline{E}\,\delta\underline{s})^T \delta\underline{u}\,d\Delta + \int_{B^e} (\underline{N}\,\delta\underline{s})^T \delta\underline{u}\,dB\right]$$

(4.3.12)

$$\left(\delta^2 T^*(\underline{s}) = \sum_e \left[\int_{\Delta^e} \delta\underline{s}^T (\delta\underline{e} - \underline{D}\,\delta\underline{u})\,d\Delta + \int_{B^e} (\underline{N}\,\delta\underline{s})^T \delta\underline{u}\,dB\right]\right).$$

Using (4.2.13) we obtain

$$\delta^2 T(\underline{s}) = \sum_e \int_{\Delta^e} \delta\underline{s}^T (\underline{D}\,\delta\underline{u})\,d\Delta$$

(4.3.13)

$$\left(\delta^2 T^*(\underline{s}) = \sum_e \int_{\Delta^e} \delta\underline{s}^T \delta\underline{e}\,d\Delta - \sum_e \int_{\Delta^e} (\underline{E}\,\delta\underline{s})^T \delta\underline{u}\,d\Delta\right)$$

and using (4.3.4)

$$\delta^2 T(\underline{s}) = \sum_e \int_{\Delta^e} \delta\underline{s}^T \delta\underline{e}\,d\Delta$$

(4.3.14)

$$\left(\delta^2 T^*(\underline{s}) = \sum_e \int_{\Delta^e} \delta\underline{s}^T \delta\underline{e}\,d\Delta\right).$$

As the material of which the structure is made is assumed stable

$$\delta \underline{s}^T \delta \underline{e} > 0 \qquad \text{if} \qquad \delta \underline{e} \neq \underline{0} . \tag{4.3.15}$$

The second variation is thus always positive and the exact solution is thus a local minimizer of $T(T^*)$. Furthermore, as the exact solution is assumed unique, it si also a proper global minimizer of $T(T^*)$.

4.4 Isoconstrained and Isominimizing Subsets

Let us consider the set X of all the structural fields a given domain Δ with boundary B decomposed into B_1 and B_2. We assume Δ subdivided into sundomains Δ^e . An internal boundary Γ is thus considered.

To each element of X corresponds a vector of strains and a vector of stresses. However, as the strains are connected to the stresses by means of (4.2.3), either the strains or the stresses are sufficient to characterize the element.

Our aim is to introduce the theory which is being described into the general frame which was presented in chapter 3. This can be done in several ways. Two possibilities are selected which seem particularly adequate to support the description of the potential energy and the complementary energy method which will be described later. The two possibilities will be presented simultaneously.

X is made a vector space by introducing the operation of addition in the following way

f_1 corresponds to $\underline{e}_1$ (f_1 corresponds to $\underline{s}_1$)

f_2 corresponds to $\underline{e}_2$ (f_2 corresponds to $\underline{s}_2$)

f_1+f_2 corresponds to $\underline{e}_1+\underline{e}_2$ (f_1+f_2 corresponds to $\underline{s}_1+\underline{s}_2$)

The set X can be made a Hilbert space X_H by associating to each pair of elements f_1 and f_2 the inner product

$$(4.4.1) \quad (f_1, f_2) = \sum_e \int_{\Delta^e} \underline{e}_1^T \underline{e}_2 \, d\Delta \qquad \left((f_1, f_2) = \sum_e \int_{\Delta^e} \underline{s}_1^T \underline{s}_2 \, d\Delta\right).$$

The norm is thus

$$(4.4.2) \quad \| f \| = \sqrt{\sum_e \int_{\Delta^e} \underline{e}^T \underline{e} \, d\Delta} \qquad \left(\| f \| = \sqrt{\sum_e \int_{\Delta^e} \underline{s}^T \underline{s} \, d\Delta}\right).$$

The isoconstrained subsets of X_H are the isocompatible (isoequilibrated) subsets of X_H. The family of functionals is the family of the total potential (complementary) energies corresponding to the different possible systems of prescribed external forces (incompatibilities) (°). Each isominimizing subset of X_H is thus an isoequilibrated (isocompatible) subset of X_H.

Each element of the space of constraints $\mathcal{G}$

(°) If the displacements prescribed on B_2 are not enough to eliminate rigid body motion, the prescribed forces must be such that equilibrium is satisfied.

corresponds to a system of incompatibilities (external forces). Each element of $\mathcal{G}$ is thus characterized by a set of initial strains $\bar{\underline{e}}^0$, displacements $\bar{\underline{u}}$ on B_2 and displacement discontinuities $\bar{\underline{g}}$ on (body forces $\bar{\underline{F}}$, tractions $\bar{\underline{p}}$ on B_1 and forces $\bar{\underline{h}}$ distributed on) Γ . The sum of two elements in $\mathcal{G}$ is characterized by the sum of the vectors $\bar{\underline{e}}^0$, $\bar{\underline{u}}$ and $\bar{\underline{g}}$ ($\bar{\underline{F}}$, $\bar{\underline{p}}$ and $\bar{\underline{h}}$) corresponding to the elements. The norm is defined by

$$\|G\| = \sum_e \int_{\Delta^e} \bar{\underline{e}}^{0T} \bar{\underline{e}}^0 \, d\Delta + \int_{B_2} \bar{\underline{u}}^T \bar{\underline{u}} \, dB + \int_{\Gamma} \bar{\underline{g}}^T \bar{\underline{g}} \, d\Gamma$$

$$\left(\|G\| = \sum_e \int_{\Delta^e} \bar{\underline{F}}^T \bar{\underline{F}} \, d\Delta + \int_{B_1} \bar{\underline{p}}^T \bar{\underline{p}} \, dB + \int_{\Gamma} \bar{\underline{h}}^T \bar{\underline{h}} \, d\Gamma \right). \qquad (4.4.3)$$

Each element of the space of functionals $\mathcal{F}$ corresponds to a system of external forces (incompatibilities). The sum of two elements in $\mathcal{F}$ is characterized by the sum of the vectors which define such magnitudes. The norm is defined by

$$\|F\| = \sum_e \int_{\Delta^e} \bar{\underline{F}}^T \bar{\underline{F}} \, d\Delta + \int_{B_1} \bar{\underline{p}}^T \bar{\underline{p}} \, dB + \int_{\Gamma} \bar{\underline{h}}^T \bar{\underline{h}} \, d\Gamma$$

$$\left(\|F\| = \sum_e \int_{\Delta^e} \bar{\underline{e}}^{0T} \bar{\underline{e}}^0 \, d\Delta + \int_{B_2} \bar{\underline{u}}^T \bar{\underline{u}} \, dB + \int \bar{\underline{g}}^T \bar{\underline{g}} \, d\Gamma \right). \qquad (4.4.4)$$

As the sum of two strain (stress) fields is compatibilized (equilibrated) by the sum of the corresponding isocompatibilities (external forces) and as the isocompatibilities (external forces) are multiplied by α if the strains (stresses) are also multiplied by α, function $g(x)$ (see Section 3.2) is linear. This means that the isoconstrained subsets of X_H are cosets of a linear subspace of X_H.

Function $f(x)$ is however generally non-linear, and the isominimizing subsets are not cosets of a linear subspace of X_H. The situation is symbolically represented in fig. 4.1.

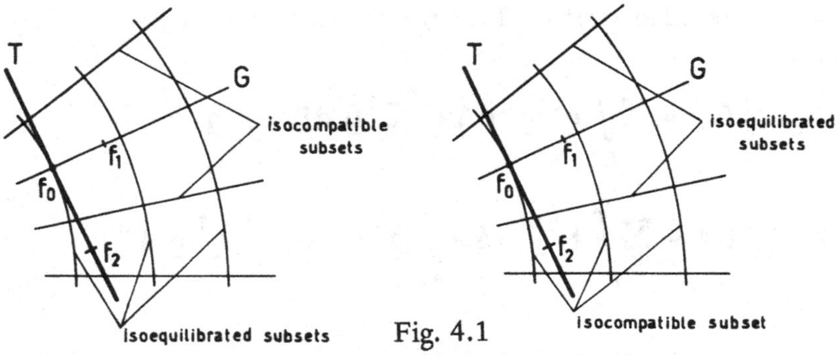

Fig. 4.1

Differentiating equation (4.2.3) ((4.2.18)), we obtain

$$\delta \underline{s} = \frac{\partial^2 W}{\partial \underline{e}\, \delta \underline{e}}\, \delta \underline{e}$$

(4.4.5)

$$\left(\delta \underline{e} = \frac{\partial^2 W^*}{\partial \underline{s}\, \partial \underline{s}}\right)$$

where

$$\frac{\partial^2 W}{\partial \underline{e}\, \partial \underline{e}} = \underline{H}(\underline{e})$$

(4.4.6)

$$\left(\frac{\partial^2 W^*}{\partial \underline{s}\, \partial \underline{s}} = \underline{H}^{-1}(\underline{s})\right)$$

represents the matrix of the second derivatives of the strain (complementary) energy density with respect to the strains (stresses).

The linear tangential manifold in a point f_0 of an isominimizing (isoconstrained) subset is the set of all the points in X_H characterized by strain (stress) vectors of the form

$$\underline{e} = \underline{e}(f_0) + \underline{H}^{-1}(f_0)\,\Delta\underline{\sigma}$$
$$\left(\underline{\sigma} = \underline{\sigma}(f_0) + \underline{H}(f_0)\,\Delta\underline{e}\right) \tag{4.4.7}$$

where $\Delta\underline{\sigma}(\Delta\underline{e})$ denotes a stress (strain) field which equilibrates (compatibilizes) vanishing external forces (incompatibilities).

Consider now the energy product

$$[f_1, f_2]_{f_0} = \sum_e \int_{\Delta^e} \underline{e}_1^T \underline{H}(f_0)\,\underline{e}_2\, d\Delta$$
$$\left([f_1, f_2]_{f_0} = \sum_e \int_{\Delta^e} \underline{\sigma}_1^T \underline{H}^{-1}(f_0)\,\underline{\sigma}_2\, d\Delta\right). \tag{4.4.8}$$

Such product is a proper inner-product because matrix $\underline{H}(f_0)$ is symmetric and positive definite.

It is easy to prove that the tangential linear manifold T in f_0 is orthogonal to the isoconstrained subset G which contains f_0. Let indeed f_1 be an arbitrary point of G. Let f_2 be an arbitrary point of T. Considering (4.4.6), we may write

$$\underline{e}_{f_2 - f_0} = \underline{H}^{-1}(f_0)\,\underline{\sigma}_{f_2 - f_0}$$
$$\left(\underline{\sigma}_{f_2 - f_0} = \underline{H}(f_0)\,\underline{e}_{f_2 - f_0}\right) \tag{4.4.9}$$

where the stresses $\underline{\sigma}_{f_2 - f_0}$ (strains $\underline{e}_{f_2 - f_0}$) are known to equilibrate (compatibilize) vanishing external forces (incompatibilities).

On the other hand, strains $\underline{e}_{f_1-f_0}$ (stresses $\underline{s}_{f_1-f_0}$) compatibilize vanishing incompatibilities (external forces). Then,

$$[f_2-f_0\,,\,f_1-f_0]_{f_0} = \sum_e \int_{\Delta^e} \underline{e}^T_{f_2-f_0}\,\underline{H}\,(f_0)\,\underline{e}_{f_1-f_0}\,d\Delta = 0 \tag{4.4.10}$$

because, by virtue of the work principle (4.2.13),

$$\sum_e \int_{\Delta^e} \underline{s}^T_{f_2-f_0}\,\underline{e}_{f_1-f_0}\,d\Delta = 0$$

$$\left(\sum_e \int_{\Delta^e} \underline{s}^T_{f_1-f_0}\,\underline{e}_{f_2-f_0}\,d\Delta = 0\right). \tag{4.4.11}$$

This means that the isoconstrained and isominimizing subsets are locally orthogonal with respect to the inner-product (4.4.8).

The definition of distance between two points can then be introduced as in Section 3.5.

4.5 The Linear Case

If the stress-strain equations (4.2.3) are linear, then matrix $\underline{H}$ does not depend on f_0 and, as geometrical linearity has been assumed, the isominimizing subsets are cosets of a linear subspace of X_H.

A new Hilbert space H can then be introduced by associating the inner-product

$$[f_1, f_2] = \sum_e \int_{\Delta^e} \underline{e}_1^T \underline{H}\, \underline{e}_2 \, d\Delta = \sum_e \int_{\Delta^e} \underline{e}_1^T \underline{s}_2 \, d\Delta = \sum_e \int_{\Delta^e} \underline{s}_1^T \underline{e}_2 \, d\Delta = \sum_e \int_{\Delta^e} \underline{s}_1^T \underline{H}^{-1} \underline{s}_2 \, d\Delta \qquad (4.5.1)$$

to each pair of elements of X_H .

The isoconstrained and isominimizing subsets are orthogonal subsets of H .

The distance between two fields becomes

$$d(f_1, f_2) = \sqrt{[f_1 - f_2, f_1 - f_2]} = \sqrt{\sum_e \int (\underline{s}_1 - \underline{s}_2)^T (\underline{e}_1 - \underline{e}_2)\, d\Delta} \qquad (4.5.2)$$

i.e. the square root of the strain energy of the difference between both,

$$d(f_1, f_2) = \sqrt{2\, U_{f_1 - f_2}} \, . \qquad (4.5.3)$$

By virtue of (4.2.23) we have still

$$d(f_1, f_2) = \sqrt{2\, U^*_{f_1 - f_2}} \, . \qquad (4.5.4)$$

The minimum total potential (complementary) energy of a field which belongs to the same isocompatible (isoequilibrated) subset as the exact solution can be then expressed in terms of the distance between f and s .

Indeed, by virtue of (4.5.1)

$$\sum_e U^e(f) = \frac{1}{2} \sum_e \int_{\Delta^e} \underline{s}^T \underline{e} \, d\Delta = \frac{1}{2} [f, f] \qquad (4.5.5a)$$

$$\left(\sum_e U^{*e}(f) = \frac{1}{2} \sum_e \int_{\Delta^e} \underline{\mathfrak{s}}^T \underline{e}\, d\Delta = \frac{1}{2}[f,f]\right). \tag{4.5.5b}$$

On the other hand, if $\underline{e}_{\mathfrak{s}}$ and $\underline{\mathfrak{s}}_{\mathfrak{s}}$ denote the strains and stresses of the exact solution, we have by virtue of (4.2.13'),

$$[\mathfrak{s},f] = \sum_e \int_{\Delta^e} \underline{\mathfrak{s}}^T_{\mathfrak{s}} \underline{e}\, d\Delta = \sum_e \int_{\Delta^e} \underline{\mathfrak{s}}^T_{\mathfrak{s}} \underline{\bar{e}}^0 d\Delta + \sum_e \int_{\Delta^e} \underline{\bar{F}}^T \underline{u}\, d\Delta + \sum_e \int_{B^e} \underline{p}^T_{\mathfrak{s}} \underline{u}\, dB$$

(4.5.6)

$$\left([f,\mathfrak{s}] = \sum_e \int_{\Delta^e} \underline{\mathfrak{s}}^T \underline{e}_{\mathfrak{s}} d\Delta = \sum_e \int_{\Delta^e} \underline{\mathfrak{s}}^T \underline{\bar{e}}^0 d\Delta + \sum_e \int_{\Delta^e} \underline{\bar{F}}^T \underline{u}_{\mathfrak{s}} d\Delta + \sum_e \int_{B^e} \underline{p}^T \underline{u}_{\mathfrak{s}}\, dB\right)$$

But

$$\sum_e \int_{B^e} \underline{p}^T_{\mathfrak{s}} \underline{u}\, dB = \int_{B_1} \underline{\bar{p}}^T \underline{u}\, dB + \int_{B_2} \underline{p}^T_{\mathfrak{s}} \underline{\bar{u}}\, dB + \int_{\Gamma} \left(\underline{p}^{+T}_{\mathfrak{s}} \underline{u}^+_{\mathfrak{s}} + \underline{p}^{-T}_{\mathfrak{s}} \underline{u}^-_{\mathfrak{s}}\right) d\Gamma$$

(4.5.7)

$$\left(\sum_e \int_{B^e} \underline{p}^T \underline{u}_{\mathfrak{s}} dB = \int_{B_1} \underline{\bar{p}}^T \underline{u}_{\mathfrak{s}} dB + \int_{B_2} \underline{p}^T \underline{\bar{u}}\, dB + \int_{\Gamma} \left(\underline{p}^{+T} \underline{u}^+_{\mathfrak{s}} + \underline{p}^{-T} \underline{u}^-_{\mathfrak{s}}\right) d\Gamma\right)$$

and

$$\begin{gathered} \underline{u}^+ - \underline{u}^- = \underline{\bar{g}} \\ \left(\underline{p}^+ + \underline{p}^- = \underline{\bar{h}}\right) \end{gathered} \tag{4.5.8}$$

$$\begin{gathered} \underline{p}^+_{\mathfrak{s}} + \underline{p}^-_{\mathfrak{s}} = \underline{\bar{h}} \\ \left(\underline{u}^+_{\mathfrak{s}} + \underline{u}^-_{\mathfrak{s}} = \underline{\bar{g}}\right) \end{gathered} \tag{4.5.9}$$

Therefore

$$\underline{p}_s^{+T}\underline{u}^+ + \underline{p}_s^{-T}\underline{u}^- = (\underline{p}_s^+ + \underline{p}_s^-)^T \frac{\underline{u}^+ - \underline{u}^-}{2} + (\underline{p}_s^+ - \underline{p}_s^-)^T \frac{\underline{u}^+ - \underline{u}^-}{2} =$$

$$= \underline{h}^T \frac{\underline{u}^+ - \underline{u}^-}{2} + \frac{1}{2}(\underline{p}_s^+ - \underline{p}_s^-)^T \underline{\bar{g}} \tag{4.5.10}$$

$$\Big(\underline{p}^{+T}\underline{u}_s^+ + \underline{p}^{-T}\underline{u}_s^- = (\underline{p}^+ + \underline{p}^-)^T \frac{\underline{u}_s^+ + \underline{u}_s^-}{2} + (\underline{p}^+ - \underline{p}^-)^T \frac{\underline{u}_s^+ - \underline{u}_s^-}{2} =$$

$$= \underline{\bar{h}}^T \frac{\underline{u}_s^+ + \underline{u}_s^-}{2} + \frac{(\underline{p}^+ - \underline{p}^-)^T}{2} \underline{\bar{g}}\Big)$$

and

$$[s, f] = \sum_e \int_{\Delta^e} \underline{s}_s^T \underline{\bar{e}}^0 d\Delta + \sum_e \int_{\Delta^e} \underline{\bar{F}}^T \underline{u}\, d\Delta + \int_{B_1} \underline{\bar{p}}^T \underline{u}\, dB + \int_{B_2} \underline{p}_s^T \underline{\bar{u}}\, dB +$$

$$+ \int_\Gamma \underline{\bar{h}}^T \frac{\underline{u}^+ + \underline{u}^-}{2} d\Gamma + \frac{1}{2}\int_\Gamma (\underline{p}_s^+ - \underline{p}_s^-)^T \underline{\bar{g}}\, d\Gamma \tag{4.5.11}$$

$$\Big([f, s] = \sum_e \int_{\Delta^e} \underline{s}^T \underline{e}_s^0 d\Delta + \sum_e \int_{\Delta^e} \underline{\bar{F}}^T \underline{u}_s d\Delta + \int_{B_1} \underline{\bar{p}}^T \underline{u}_s dB + \int_{B_2} \underline{p}^T \underline{\bar{u}}\, dB +$$

$$+ \int_\Gamma \underline{\bar{h}}^T \frac{\underline{u}_s^+ + \underline{u}_s^-}{2} d\Gamma + \int_\Gamma \frac{(\underline{p}^+ - \underline{p}^-)^T}{2} \underline{\bar{g}}\, d\Gamma\Big)$$

Considering (4.3.1) we can write

$$T(f) = \frac{1}{2}[f, f] - [s, f] + \sum_e \int_{\Delta^e} \underline{s}_s^T \underline{\bar{e}}^0 d\Delta + \int_{B_2} \underline{p}_s^T \underline{\bar{u}}\, dB +$$

$$+ \int_\Gamma \frac{(\underline{p}_s^+ - \underline{p}_s^-)^T}{2} \underline{\bar{g}}\, d\Gamma \tag{4.5.12}$$

$$\Big(T^*(f) = \frac{1}{2}[f, f] - [f, s] + \sum_e \int_{\Delta^e} \underline{\bar{F}}^T \underline{u}_s d\Delta + \int_{B_1} \underline{\bar{p}}^T \underline{u}_s dB +$$

$$+ \int_\Gamma \underline{\bar{h}}^T \frac{\underline{u}_s^+ + \underline{u}_s^-}{2} d\Gamma\Big)$$

Particularizing for $s = f$, we obtain

$$T(s) = \frac{1}{2}[s,s] - [s,s] + \sum_e \int_{\Delta^e} \underline{s}_s^T \underline{\bar{e}}^0 \, d\Delta + \int_{B_2} \underline{p}_s^T \underline{\bar{u}} \, dB + \int_{\Gamma} \frac{1}{2}(\underline{p}_s^+ - \underline{p}_s^-)^T \underline{\bar{g}} \, d\Gamma$$

(4.5.13)

$$\left(T^*(s) = \frac{1}{2}[s,s] - [s,s] + \sum_e \int_{\Delta^e} \underline{\bar{f}}^T \underline{\bar{u}} \, d\Delta + \int_{B_1} \underline{\bar{p}}^T \underline{u}_s \, dB + \int_{\Gamma} \underline{\bar{h}}^T \frac{\underline{u}_s^+ + \underline{u}_s^-}{2} \, d\Gamma\right)$$

Subtracting (4.5.13) from (4.5.12), there results

$$T(f) - T(s) = \frac{1}{2}[f,f] - [s,f] + \frac{1}{2}[s,s]$$

(4.5.14)

$$\left(T^*(f) - T^*(s) = \frac{1}{2}[f,f] - [s,f] + \frac{1}{2}[s,s]\right)$$

or

$$T(f) - T(s) = \frac{1}{2}[f-s, f-s] = \frac{1}{2} d^2[f,s]$$

(4.5.15)

$$\left(T^*(f) - T^*(s) = \frac{1}{2}[f-s, f-s] = \frac{1}{2} d^2[f,s]\right)$$

The total potential (complementary) energy can thus be given the following form:

"In any class of compatible (equilibrated) fields, the same field which minimizes the total potential (complementary) energy minimizes the distance to the exact solution".

The theorem can be stated in even more general terms: "In any class of isocompatible (isoequilibrated) fields the same field which minimizes the total potential (complementary) energy minimizes the distance to the exact solution".

Let indeed f minimize T on a subset of the

isocompatible subset C_1 (T^* on a subset of the isoequilibrated subset E_1) see fig. 4.2). Then, as Pythagora's theorem holds,

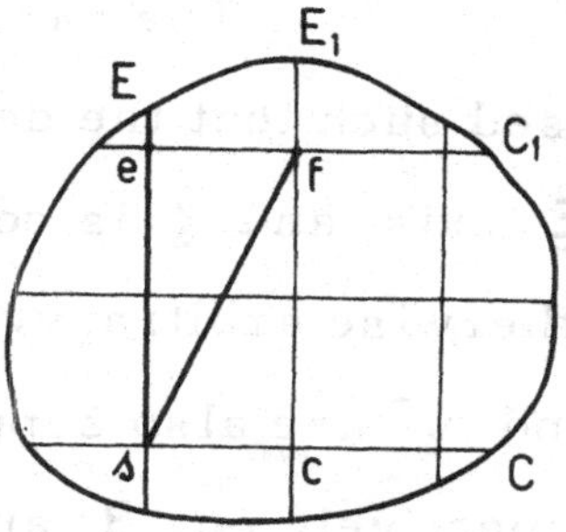

Fig. 4.2

$$d^2(s, f) = d^2(e, f) + d^2(e, s) \tag{4.5.16}$$

$$\left(d^2(s, f) = d^2(c, f) + d^2(c, s)\right)$$

and, as $d(e, f)$ ($d(e,f)$) is minimum and $d(e,s)$ ($d(c,s)$) is fixed, $d(s, f)$ is also minimum.

4.6 The Variational Theorems

The total potential and complementary energy theorems will be introduced now in the general frame of the variational theorems [23] . The terms which concern the internal boundary Γ will be omitted in the expressions of the functionals with the aim of simplifying the equations.

Let us consider the total potential energy theorem and use the Lagrange multiplier technique to get free of the constraints (4.2.2), (4.2.9) and (4.2.11). The functional which becomes stationary is:

$$\Phi = U - \int_\Delta \underline{\bar{F}}^T \underline{u}\, d\Delta - \int_\Delta \underline{s}^T (\underline{e} - \underline{\bar{e}}^0 - \underline{D}\,\underline{u})\, d\Delta - \int_{B_1} \underline{\bar{p}}^T \underline{u}\, dB - \int_{B_2} \underline{p}^T (\underline{u} - \underline{\bar{u}})\, dB \tag{4.6.1}$$

The variations of vector $\underline{u}$, $\underline{\sigma}$ and $\underline{e}$ are supposed such that the derivatives involved in operators $\underline{D}$ and $\underline{E}$ exist and $\underline{\sigma}$ is connected to $\underline{e}$ through (4.2.3) (°) but are otherwise arbitrary. Vectors $\underline{p}$ and $\underline{u}^+$, and $\underline{p}^+$, $\underline{p}^-$, $\underline{u}^+$ and $\underline{u}^-$ are also supposed to vary independently of each other respectively on B and Γ . Equation (4.2.4) is however assumed to hold (°). The exact solution makes Φ stationary.

Reissner's theorem [24] is a particular case of the one which has been established. It assumes equation (4.2.2) to be satisfied and suppose that no internal boundary Γ exists. The functional is thus

$$\Phi = U - \int_{\Delta} \underline{\bar{F}}^T \underline{u}\, d\Delta - \int_{B_1} \underline{\bar{p}}^T \underline{u}\, dB - \int_{B_2} \underline{p}^T (\underline{u} - \underline{\bar{u}})\, dB \tag{4.6.2}$$

or considering (4.2.16) and (4.2.2) (with $\underline{e}^0 = \underline{0}$),

$$\Phi = \int_{\Delta} \left[\underline{\sigma}^T (\underline{D}\,\underline{u}) - \underline{\bar{F}}^T \underline{u} - W\right] d\Delta - \int_{B_1} \underline{\bar{p}}^T \underline{u}\, dB - \int_{B_2} \underline{p}^T (\underline{u} - \underline{\bar{u}})\, dB \tag{4.6.3}$$

The total complementary energy theorem is also contained in the general theorem. Assuming indeed equations (4.2.1), (4.2.8) and (4.2.12) to be satisfied, functional F takes the form

---

(°) Of course conditions (4.2.3) and (4.2.4) could also be omitted. Such degree of generality is not needed however.

$$\Phi = -U^* + \int_\Delta \left[\underline{s}^T(\underline{D}\,\underline{u}) - (\underline{E}\,\underline{s})^T\underline{u}\right] d\Delta + \int_\Delta \underline{s}^T\underline{\bar{e}}^0\, d\Delta -$$

$$- \int_{B_1} \underline{p}^T\underline{u}\, dB - \int_{B_2} \underline{p}^T(\underline{u} - \underline{\bar{u}})\, dB. \tag{4.6.4}$$

By virtue of (4.2.13),

$$\int_\Delta \left[\underline{s}^T(\underline{D}\,\underline{u}) - (\underline{E}\,\underline{s})^T\underline{u}\right] d\Delta = \int_B \underline{p}^T\underline{u}\, dB. \tag{4.6.5}$$

Introducing (4.6.5) in (4.6.4), we obtain

$$\Phi = -\left[U^* - \int_\Delta \underline{s}^T\underline{\bar{e}}^0\, d\Delta - \int_{B_2} \underline{p}^T\underline{\bar{u}}\, dB\right] = -T^*. \tag{4.6.6}$$

Now, as the exact solution minimizes the total complementary energy on the set of the equilibrated fields, and

$$\Phi = -T^* \tag{4.6.7}$$

we conclude that the exact solution maximizes functional Φ on the set of the compatible fields.

Functional Φ, whose expression is given by (4.6.1), is thus minimized by s on the isocompatible subset C, which contains the exact solution, and maximized by s on the isoequilibrated subset E which also

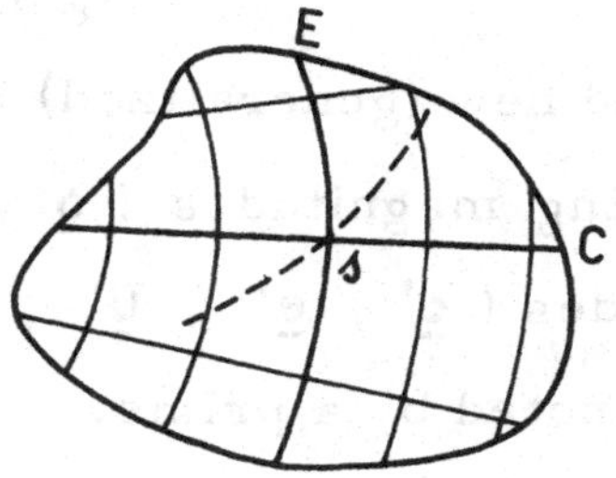

Fig. 4.3

contains the exact solution (fig. 4.3).

Considering a two-dimensional symbolic representation of space X_H , (fig. 4.3) and associating to each point of X_H the corresponding value of functional Φ , we define a surface with a stationary point at s . As s is a minimizer on C and a maximizer on E , the surface has the shape of a saddle. This explains why the variational theorem associated to functional Φ expressed by (4.6.1) is no longer a minimum theorem.

4.7 The Generation Methods

Two methods will be presented in the sequel for the generation of new analytical models starting from an initial one. The two methods will be considered simultaneously in connexion with the general descriptions also simultaneously introduced in Section 4.4.

Magnitudes referring to the initial (generating) and new (generalized) model will be called respectively generating magnitudes ($\underline{s}$, $\underline{e}$, $\underline{u}$, ...) and generalized magnitudes ($\underline{s}'$, $\underline{e}'$, $\underline{u}'$, ...). Generalized magnitudes will be denoted by a prime.

It is assumed that the generating and generalized models refer to domains Δ and Δ' , i.e. that the generating magnitudes distribute on Δ and the generalized magnitudes on Δ' .

Δ' is assumed to be a submanifold of Δ . It is also assumed that $d\Delta$ can be expressed as the product of $d\Delta'$ by another differential, $d\Delta''$, i.e.

$$d\Delta = d\Delta \cdot d\Delta'' \tag{4.7.1}$$

and that the integrals on Δ can be transformed into integrals on Δ' by

$$\int_{\Delta} (\ ) \, d\Delta = \int_{\Delta'} \left[\int_{\Delta''_t} (\ ) \, d\Delta'' \right] d\Delta' \tag{4.7.2}$$

in which Δ''_t denotes a family of subdomains (°) of Δ whose parameter, to, represents any point in Δ' .

Given a subdomain $\Delta^{e'} \subset \Delta'$ the corresponding subdomain $\Delta^e \subset \Delta$ is defined as the union of all the subdomains Δ'_t corresponding to every point in $\Delta^{e'}$, i.e.

$$\Delta^e = \bigcup_{t \in \Delta^{e'}} \Delta''_t \tag{4.7.3}$$

(°) If the generating model is the three-dimensional one and the generalized model is the theory of shells, for instance, Δ' is the middle surface of the shell and Δ''_t is the portion of the normal to the middle surface contained within the thickness t of the shell.
Integrations on Δ''_t are thus integrations between $-t/2$ and $t/2$.

Integrals on Δ^e can thus be transformed into integrals on $\Delta^{e'}$ by

(4.7.4) $$\int_{\Delta^e} (\ ) \, d\Delta = \int_{\Delta^{e'}} \left[\int_{\Delta''_t} (\ ) \, d\Delta'' \right] d\Delta'$$

The strain energy associated with subdomain Δ^e may thus be expressed by (4.2.14) or by

(4.7.5) $$U = \int_{\Delta^{e'}} W' \, d\Delta'$$

where

(4.7.6) $$W' = \int_{\Delta''_t} W \, d\Delta'$$

is the density with respect to Δ'.

The boundary $B^{e'}$ of $\Delta^{e'}$ is the intersection of Δ' with the boundary B^e of Δ^e. The boundary B' of Δ' is the intersection of Δ' with boundary B of Δ. B'_1 and B'_2 denote the intersection of Δ' with B and B_2. Γ' denotes the intersection of Δ with Γ.

It is admitted that integrals on B^e can be transformed into integrals on $B^{e'}$ by

(4.7.7) $$\int_{B^e} (\ ) \, dB = \int_{B^{e'}} \left[\int_{\Delta''_t} (\ ) \, d\Delta'' \right] dB'$$

and that integrals on B, B_1, B_2 and Γ can similarly be transformed.

A last assumption is that, by application of the

divergence theorem, the identity

$$\int_{\Delta^{e'}} \underline{\xi}^T (\underline{D}'\underline{\eta})\, d\Delta' = \int_{\Delta^{e'}} (\underline{E}'\underline{\xi})^T \underline{\eta}\, d\Delta' + \int_{B^{e'}} (\underline{N}'\underline{\xi})^T \underline{\eta}\, dB' \quad (4.7.8)$$

can be established for any subdomain $\Delta^{e'} \subset \Delta'$ with boundary $B^{e'}$, and for any pair of vectors $\underline{\xi}$ and $\underline{\eta}$ defined on $\Delta^{e'}$, such that $\underline{\xi}$ has a number of elements equal to the number of columns in $\underline{E}'$ and $\underline{\eta}$ a number of elements equal to the number of columns in $\underline{D}'$. It is also assumed that the elements of $\underline{\xi}$ and $\underline{\eta}$ admit the derivatives involved in $\underline{E}'$ and $\underline{D}'$ and are otherwise arbitrary.

Identity (4.7.8) must not be regarded as having any special physical meaning. Operators $\underline{E}'$, $\underline{D}'$ and $\underline{N}'$ are arbitrary. The important point is that given an operator $\underline{D}'$, it is always possible, by using the divergence theorem, to expand $\int_{\Delta^{e'}} \underline{\xi}^T(\underline{D}'\underline{\eta})\, d\Delta'$ in two integrals, one on $\Delta^{e'}$, and the other on $B^{e'}$, and thus determine operators $\underline{E}'$ and $\underline{N}'$. Also given an operator $\underline{E}'$, it is possible to expand $\int_{\Delta^{e'}} (\underline{E}'\underline{\xi})^T \underline{\eta}\, d\Delta'$ and thus determine $\underline{D}'$ and $\underline{N}'$.

The two generation methods which will be described are the potential energy method and the complementary energy method.

The potential (complementary) energy method consists of the following steps:

a) choice of a set of compatibility (equilibrium) conditions of the generating model for generalized strain-displacement (equilibrium) equations. Such conditions have the form (°)

$$\underline{e}' = \underline{D}'\underline{u}' + \underline{e}^{o'}$$

(4.7.9)

$$\left(\underline{E}'\underline{s}' = \underline{f}' \qquad \underline{N}'\underline{s}' = \underline{p}'\right)$$

in which $\underline{e}'$ and $\underline{u}'$ ($\underline{s}'$ and $\underline{p}'$) are generally functions of the generating strains and displacements (stresses and tractions) which distribute on Δ' and are chosen for generalized strains and displacements (stresses and tractions).

The corresponding generalized-generating relations

(4.7.10)

$$\underline{u}' = \underline{\alpha}(\underline{u}) \qquad \left(\underline{s}' = \underline{\nu}(\underline{s})\right)$$

$$\underline{e}' = \underline{\beta}(\underline{e}) \qquad \left(\underline{p}' = \underline{\omega}(\underline{p})\right)$$

result from the equations themselves. The same happens to the initial strains and prescribed displacements (external force densities) and to their generalized-generating relations

(4.7.11) $$\bar{\underline{e}}^{o'} = \bar{\underline{e}}^{o'}(\underline{e}^{o}) \qquad \bar{\underline{f}}' = \bar{\underline{f}}'(\bar{\underline{f}})$$

(4.7.12) $$\bar{\underline{u}}' = \bar{\underline{u}}'(\bar{\underline{u}}) \qquad \bar{\underline{p}}' = \bar{\underline{p}}'(\bar{\underline{p}})$$

(°) Equations in the left (right) respect the first (second) method.

Generalized-generating relations

$$\bar{\underline{\Delta}}' = \bar{\underline{\Delta}}(\bar{\underline{\Delta}}) \qquad\qquad \bar{\underline{\pi}}' = \bar{\underline{\pi}}'(\bar{\underline{\pi}}) \qquad (4.7.13)$$

must be introduced for the displacement discontinuities (external forces) prescribed on Γ .

The strain-displacement (equilibrium) equations (4.7.1) are exact within the framework of the generating theory. This is a very important point which will be commented later and which is responsible for compatibility (equilibrium) being preserved in the generating-generalized sense.

The simplifying assumption which is implicit at this stage is that the generating displacements and strains (stresses and tractions) can be uniquely expressed in terms of the corresponding generalized magnitudes. The definition of the generating-generalized displacement and strain (stress and traction) relations is however not yet required.

b) Definition of the generalized stresses (strains) as derivatives of the generalized strain (complementary) energy density (°)

(°) The assumption of the strain (complementary) energy being an invariant is implicit in (4.7.6).

(4.7.14)
$$W'(\underline{e}') = \int_{\Delta''_t} W\left[\underline{e}(\underline{e}')\right] d\Delta''$$
$$\left(W^{*'}(\underline{s}') = \int_{\Delta''_t} W^{*}\left[\underline{s}(\underline{s}')\right] d\Delta''\right)$$

with respect to the generalized strains (stresses), i.e.

(4.7.15)
$$\underline{s}' = \frac{\partial W}{\partial \underline{e}'} \qquad \left(\underline{e}' = \frac{\partial W^{*}}{\partial \underline{s}'}\right)$$

c) Determination of the generalized equilibrium (strain-displacement) equations by assuming the work principle to hold in the generalized model, i.e.,

(4.7.16)
$$\int_{\Delta^{e'}} \underline{s}'^{T} \underline{e}' d\Delta' = \int_{\Delta^{e'}} \underline{s}'^{T} \underline{e}^{0'} d\Delta' + \int_{\Delta^{e'}} \underline{f}'^{T} \underline{u}' d\Delta' + \int_{B^{e'}} \underline{p}'^{T} \underline{u}' dB'$$

in which the generalized stresses and displacements are assumed to admit within the domain $\Delta^{e'} \subset \Delta'$, the derivatives involved in the operators D' and E' .
By using (4.7.8) with

(4.7.17)
$$\xi = s'$$

(4.7.18)
$$\eta = u'$$

(4.7.16) can be transformed into

$$\int_{\Delta^{e'}} (\underline{E}'\underline{s}' - \underline{f}')^T \underline{u}' \, d\Delta' + \int_{B^{e'}} (\underline{N}'\underline{s}' - \underline{p}')^T \underline{u}' \, dB' = 0$$

$$\left(\int_{\Delta^{e'}} \underline{s}'^T (\underline{e}' - \underline{e}^{o'} - \underline{D}'\underline{u}') \, d\Delta' = 0 \right) \qquad (4.7.19)$$

As $\underline{u}'(\underline{s}')$ is arbitrary, equations

$$\underline{E}'\underline{s}' = \underline{f}' \qquad \qquad (\underline{e}' = \underline{e}^{o'} + \underline{D}'\underline{u}') \qquad (4.7.20)$$
$$\underline{N}'\underline{s}' = \underline{p}'$$

result, which are taken for generalized equilibrium (strain-displacement) equations.

d) Explicitation of the generating-generalized strain (stress) relations

$$\underline{e} = \underline{\chi}\,\underline{e}' \qquad \qquad (\underline{s} = \underline{\psi}\,\underline{s}') \qquad (4.7.21)$$

Matrices $\underline{\chi}$ and $\underline{\psi}$ must be such that

$$\underline{e}' = \underline{\beta}\ \underline{\chi}\,\underline{e}' \qquad \qquad (\underline{s}' = \underline{\nu}\,(\underline{\psi}\,\underline{s}')) \qquad (4.7.22)$$

in order that (4.7.10) be satisfied.

The strain (complementary) energy can now be determined as a function of the generalized strains (stresses) and the stress-strain (strain-stress) equations be determined by developing (4.7.15).

Introducing (4.2.19) ((4.2.21)) in (4.7.14), we obtain

$$2\,W'(\underline{e}') = \int_{\Delta''_t} \underline{e}^T \underline{H}\, \underline{e}\, d\Delta'' = \underline{e}'^T \int_{\Delta''_t} \underline{\chi}^T \underline{H}\, \underline{\chi}\, d\Delta''\, \underline{e}' = \underline{e}'^T \underline{H}' \underline{e}'$$

(4.7.23)

$$\left(2\,W^{*\prime}(\underline{s}') = \int_{\Delta''_t} \underline{s}^T \underline{H}^{-1} \underline{s}\, d\Delta'' = \underline{s}'^T \int_{\Delta''_t} \underline{\psi}^T \underline{H}^{-1} \underline{\psi}\, d\Delta''\, \underline{s}' = \underline{s}'^T \underline{H}'^{-1} \underline{s}'\right)$$

in which

(4.7.24)

$$\underline{H}' = \int_{\Delta''_t} \underline{\chi}^T \underline{H}\, \underline{\chi}\, d\Delta''$$

$$\underline{H}'^{-1} = \int_{\Delta''_t} \underline{\psi}^T \underline{H}^{-1} \underline{\psi}\, d\Delta''$$

The generalized stress strain (strain-stress) equations

(4.7.25) $$\underline{s}' = \underline{H}' \underline{e}' \qquad \left(\underline{e}' = \underline{H}'^{-1} \underline{s}'\right)$$

can be obtained by introducing (4.7.23) in (4.5.15).

e) Introduction of the generating-generalized displacement (traction) relations,

(4.7.26) $$\underline{u} = \underline{\varphi}\, \underline{u}' \qquad \left(\underline{p} = \underline{\gamma}\, \underline{p}'\right)$$

Matrices $\underline{\varphi}$ and $\underline{\gamma}$ must be such that

(4.7.27) $$\underline{u}' = \underline{\alpha}\left(\underline{\varphi}\, \underline{u}'\right) \qquad \underline{p}' = \underline{\omega}\left(\underline{\gamma}\, \underline{p}'\right)$$

The introduction of relations (4.7.26) and the assumption of the total potential (complementary) energy being an invariant lead to the generalized-generating external force (incompatibility) relations.

Indeed, as the strain (complementary) energy is an invariant itself, the total potential (complementary) energy will be an invariant if

$$\int_{\Delta^e} \bar{\underline{F}}^T \underline{u}\, d\Delta = \int_{\Delta^{e'}} \underline{F}'^T \underline{u}'\, d\Delta \qquad \left(\int_{\Delta^e} \underline{s}^T \bar{\underline{e}}^0 d\Delta = \int_{\Delta^e} \underline{s}'^T \bar{\underline{e}}^{0'} d\Delta \right) \qquad (4.7.28)$$

$$\int_{B_1} \bar{\underline{p}}^T \underline{u}\, d\Delta = \int_{B_1} \bar{\underline{p}}'^T \underline{u}'\, d\Delta \qquad \left(\int_{B_2} \underline{p}^T \bar{\underline{u}}\, dB = \int_{B_2} \underline{p}'^T \bar{\underline{u}}'\, dB \right) \qquad (4.7.29)$$

$$\int_{\Gamma} \bar{\underline{h}}^T (\underline{u}^+ + \underline{u}^-)\, d\Gamma = \int_{\Gamma'} \bar{\underline{h}}'^T (\underline{u}'^- - \underline{u}'^+)\, d\Gamma' \qquad \left(\int_{\Gamma} (\underline{p}^+ + \underline{p}^-)^T \bar{\underline{g}}\, d\Gamma = \int_{\Gamma'} (\underline{p}'^+ + \underline{p}'^-)^T \bar{\underline{g}}'\, d\Gamma \right) \qquad (4.7.30)$$

In (4.7.28), $\underline{u}'$ ($\underline{s}'$) is any continuous generalized displacement (stress) field defined on an arbitrary subdomain $\Delta^e \subset \Delta'$ and $\underline{u}$ ($\underline{s}$) is the equivalent generating displacement (stress) field defined on the corresponding subdomain, $\Delta^e \subset \Delta$. In (4.7.29) and (4.7.30), $\underline{u}'$ ($\underline{p}'$) is any continuous displacement (traction) field respectively on B_2' and Γ'. $\underline{u}$ ($\underline{p}$) is the equivalent generating field defined on B_2 and Γ .

Introducing (4.7.21) and (4.7.26) in (4.7.28), (4.7.29) and (4.7.30) the following relations result

$$\bar{\underline{F}}' = \int_{\Delta_t''} \underline{\varphi}^T \bar{\underline{F}}\, d\Delta'' \qquad \bar{\underline{e}}^{0'} = \int_{\Delta_t''} \underline{\psi}^T \bar{\underline{e}}^0 d\Delta'' \qquad (4.7.31)$$

$$\bar{\underline{p}}' = \int_{\Delta_t''} \underline{\varphi}^T \bar{\underline{p}}\, d\Delta'' \qquad \bar{\underline{u}}' = \int_{\Delta_t''} \underline{\gamma}^T \bar{\underline{u}}\, d\Delta'' \qquad (4.7.32)$$

$$\bar{\underline{h}}' = \int_{\Delta_t''} \underline{\varphi}^T \bar{\underline{h}}\, d\Delta'' \qquad \bar{\underline{g}}' = \int_{\Delta_t''} \underline{\gamma}^T \bar{\underline{g}}\, d\Delta'' \qquad (4.7.33)$$

4.8 Interpretation of the Generation Methods

The generation methods will be interpreted now in terms of the general frame introduced in Section 4.4. It is important to recognize that the two methods are distinct.

Their most important distinctions consist in this:

a) The potential (complementary) energy is invariant in the first (second) method.

b) In the first (second) method, compatibility (equilibrium) is preserved in the generating-generalized sense.

In what concerns distinction b), we remember again that concepts such as compatibility, equilibrium and exact solution are relative concepts both in the generating and in the generalized theory.

Compatibility must indeed be related to a set of vectors $\bar{\underline{e}}^0$, $\bar{\underline{u}}$ and $\bar{\underline{g}}$, and equilibrium to a set of vectors $\bar{\underline{f}}$ $\bar{\underline{p}}$ and $\bar{\underline{h}}$. The exact solution has no meaning valueless vectors $\bar{\underline{e}}^0, \bar{\underline{u}}, \bar{\underline{g}}, \bar{\underline{f}}, \bar{\underline{p}}$ and $\bar{\underline{h}}$ have been prescribed. Generalized compatibility and equilibrium, as well as the generalized exact solution, are connected to the corresponding generating concepts because, equations (4.7.3) and (4.7.4) permit to obtain the generalized vectors $\bar{e}^{0'}$, ... in terms of the corresponding generating vectors $\bar{e}^0$, ...

As in the potential (complementary) energy method, the generalized compatibility (equilibrium) conditions are obtained by particularizing the generating ones, it follows that any generating compatible (equilibrated) field generates, through (4.7.2) a generalized compatible (equilibrated) field. This is what is meant by stating that compatibility (equilibrium) is preserved in the generating-generalized sense.

However it must be recognized that neither compatibility nor equilibrium are necessarily preserved in the generalized-generating sense, i.e. that the fact that a generalized field is compatible or equilibrated in the generalized way does not imply that the corresponding generating field (corresponding through (4.7.13) and (4.7.18)) be compatible or equilibrated in the generating way.

The generalized-generating relations (4.7.2) introduce a correspondence between elements in X_H and elements in the space of the generalized fields X_H^G . To each element $f \in X_H$ there corresponds a single element $f^G \in X_H^G$ but to each element $f^G \in X_H^G$ there corresponds a subset S_f of X_H . In other words, the space X_H is decomposed into classes of equivalence and to each of such classes a unique element $f^G \in X_H^G$ corresponds.

The generating-generalized relations (4.7.13) and (4.7.18) introduced a one-to-one correspondence between each element $f^G \in X_H^G$ and an element f in X_H, or a correspondence between each class of equivalence S_f and one of its elements, f' (fig. 4.4). Such element f' is the T -image of every element f of S_f .

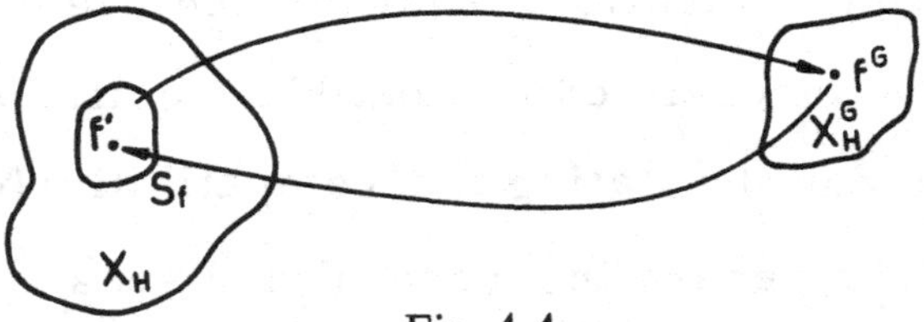

Fig. 4.4

As the strain (complementary) energy was made an invariant, the distance between two generalized fields in X_H^G is equal to the distance between the two corresponding generating fields in X_H . The range of the operator T is thus isometric of X_H^G .

Furthermore, as the total potential (complementary) energy was made an invariant, and the minimum theorems subsist in the generalized model, the same functionals which are minimized on X_H are minimized on X'_H .

On the other hand, the T -images of isoconstrained elements in X_H are isoconstrained elements in X'_H . This means that in the potential (complementary) energy method, isocompatible (isoequilibrated) elements in X_H generate isocompatible (isoequilibrated) elements in X_H^G , and therefore, compatibility (equilibrium) is preserved in the generating-generalized sense. However, only if the operator T is conforming, compatibility (equilibrium) will be preserved in the generalized-generating sense.

4.9 Justification of the Generation Methods

The generation methods which have been described lead to generalized models which are perfectly analogous to the generating ones. The equilibrium, strain-displacement and stress-strain equations, together with the work principle and the positive-definiteness of matrix $\underset{\sim}{H}$ are indeed sufficient for establishing every general theorem in the theory of structures. Particularly the variational theorems, including the minimum theorems are valid both in the generalized and in the generating models. The generalized exact solution minimizes thus the generalized total potential energy in the set of the generalized compatible fields and the total complementary energy in the set of the generalized equilibrated fields.

This great theoretical advantage is not however a justification in itself, if one considers both methods under the point of view of efficiency. It must not be forgotten indeed that the structural models are conceived for the analysis of structures, and they are to be rejected if they do not provide reasonably accurate results, i.e. results which are sufficiently close to the ones provided by the three-dimensional model, which generates all the others.

Each generalized model is thus an approximation to the corresponding generating one, and the solution it provides, i. e. the generalized exact solution, is an approximation to the generating exact solution.

A justification for the potential (complementary) energy method can easily be given if compatibility (equilibrium) is preserved in the generalized-generating sense and if linearity is assumed.

Indeed, then, the approximate solution s' which corresponds to the generalized solution minimizes the total potential (complementary) energy on a set of generating compatible (equilibrated) fields, and thus, by virtue of the special form of the total potential (complementary) energy theorem valid for the linear case (see Section 4.5) the approximate solution is the field of such set which is nearer to the generating exact solution, s.
On the other hand, under the same conditions,

$$T(s) \leq T(s'_a) \qquad (4.9.1)$$
$$T^*(s) \leq T^*(s'_a)$$

and this provides bounds for the strain (complementary) energy. Assume that all prescribed external forces (incompatibilities) vanish. Then (see (4.3.1)),

$$T = U \qquad (4.9.2)$$
$$T^* = U^*$$

and thus, by virtue of (4.9.1),

$$U(s) \leq U(s'_a)$$
$$\left(U^*(s) \leq U^*(s)\right) \tag{4.9.3}$$

We remark that (4.9.3) is valid even in the non-linear case.

Assume now that the prescribed isocompatibilities (external forces) vanish. Then in the linear case, by virtue of (4.5.13)

$$T(s) = -U(s)$$
$$\left(T^*(s) = -U^*(s)\right) \tag{4.9.4}$$

and, as equality (4.5.13) is still valid for the generalized model,

$$T(s'_a) = -U'(s'_a)$$
$$\left(T^*(s'_a) = -U^*(s'_a)\right) \tag{4.9.5}$$

Hence, by virtue of (4.9.1)

$$U(s) \geq U(s'_a)$$
$$\left(U^*(s) \geq U^*(s'_a)\right) \tag{4.9.6}$$

In general, however, compatibility (equilibrium) is not preserved in the generalized-generating senses and the justification of the generation methods must be searched elsewhere.

4.10 The Role of Convergence

The question is that the problem of the justification of the generation methods in the theory of structures is only a particular case of the general problem of the justification of approximate methods in mathematical analysis and the most general criterion used in mathematical analysis is the convergence criterion according to which an approximate method is to be rejected if cannot lead to a sequence of approximate solutions converging to the exact one.

In the case of the finite element method, for instance, such sequence can be established by using successive subdivisions of the domain Δ into subdomains with decreasing size.

Convergence can appear however, in the theory of structures still in a different situation. Such is the case in the theory of shells, whose justification depends on the exact solution provided by three-dimensional elasticity becoming closer to the solution obtained from the theory of shells as the thickness of the shell tends to zero.

Now, the generalized solution provided by the theory of shells does not depend on the thickness, provided the two-dimensional elastic coefficients (see Section 6.4) are kept constant when the thickness tends to zero. However, we remark that this condition cannot be respected in an ordinary shell because, as the bending coefficients depend on t^3 and the membrane coefficients on t, they cannot be kept all constant as t tends to zero. Such difficulty disappears only in a generalized shell, i.e. in a shell where the couple-stresses are not assumed to vanish.

The first situation is illustrated in fig. 4.5. Each subdivision into finite elements generates a generalized model characterized by a certain number of generalized displacements, strains and stresses. To each generalized model corresponds a space of generalized fields $X^G_{H_n}$ which is associated to a subspace $X'_{H_n} \subset X_H$ and an operator T_n.

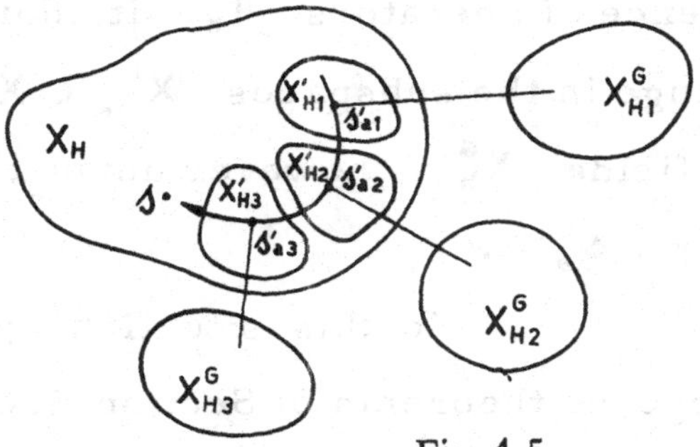

Fig. 4.5

We can apply then the first pair of convergence theorems given in Section 3.6. Such application is the base for the discussion of convergence in the finite element method which will be presented in chapter 7.

We remark that the projection theorem can be used if the definition of distance is introduced as in section 3.5 (see sect. 4.4). The second situation

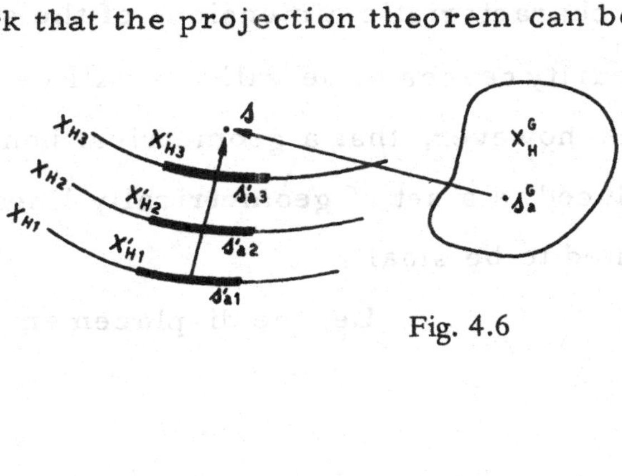

Fig. 4.6

is illustrated in fig. 4.6.

Each value of the thickness of the shell leads to a domain Δ_n and to a space X_{H_n}. We have thus a sequence of subspaces X_{H_n}, a sequence of spaces X'_{H_n} and a sequence of operators T_n with domain in the space X_{H_n} and range in the subspaces $X'_{H_n} \subset X_{H_n}$. The space of generalized fields X_H^G is however unique, and so is the generalized solution s_a^G.

To this situation applies the second pair of convergence theorems of Section 3.6.

Although the theory of shells will be presented in chapter 6, convergence will not be discussed there, and therefore such theorems will not be applied. A paper about this point is intended to be published in a near future.

4.11 Geometrical Non-Linear Cases

If the displacements are large with respect to the characteristic dimensions of the structure, geometrical linearity ceases to be valid. It will be shown in the present section, however, that a geometrical non-linear problem can be reduced to a set of geometrically linear ones. Strains are assumed to be small.

Let the displacement vector $\underline{u}$ be decomposed

into two terms

$$\underline{u} = \underline{u}_n + \underline{u}_s \qquad (4.11.1)$$

The first term represents a rigid body motion of the structure, the second term a small displacement.

As the strains are due only to the second term we have

$$\underline{e} = \underline{D}\,\underline{u}_s + \underline{e}^0 = \underline{D}\left(\underline{u} - \underline{u}_n\right) + \underline{e}^0 \qquad (4.11.2)$$

Introducing (4.11.2) in (4.2.20), and then (4.2.20) in (4.2.1), Navier's equations (4.2.24) are replaced by

$$\underline{E}\,\underline{H}\,\underline{D}\,\underline{u} = \underline{f} + \underline{E}\,\underline{H}\,\underline{D}\,\underline{u}_n \qquad (4.11.3)$$

and the stress boundary conditions by

$$\underline{N}\,\underline{H}\,\underline{D}\,\underline{u} = \underline{\bar{p}} + \underline{N}\,\underline{H}\,\underline{D}\,\underline{u}_n \qquad (4.11.4)$$

These equations are of course written for the deformed configuration.

We remark that $(\underline{D}\,\underline{u}_n)$ does not vanish, although $\underline{u}_n$ denotes a rigid body motion because the rotations associated to $\underline{u}_n$ are not small. If they are, then $\underline{D}\,\underline{u}_n$ becomes neglectable.

In practice equations (4.11.3) and (4.11.4) can be used with an incremental procedure: at each step vector can be determined by considering the rotations of the structure in the preceding step.

The theory which was presented along this chapter can be however generalized to cover geometrical non-linearities. It suffices to remember that the total potential energy theorem is valid even in that case. Under a computational point of view however, the reduction to the linear case as it is done in the present section seems the best way to deal with such problems. If applied to the discrete model namely, it makes unnecessary the consideration of the geometrical stiffness matrix which usually is employed.

5. THREE-DIMENSIONAL MODEL

5.1 Introduction

The classical three-dimensional theory of elasticity is generally thought of as the only one available to describe the three-dimensional behaviour of elastic bodies. The classical theory is however nothing else than a particular case of the much more general theories of the so-called generalized continua [25].

The basic concepts of such theories can be traced back to MacCullagh, Cauchy, Kelvin, Voigt and Duhem, but the first systematic presentation was due to the brothers Cosserat [26] (1909), whose work was almost unnoticed at the time of publication. The idea however was rediscovered fifty years later and fully developed since then.

The basic physical model of the generalized continua consists of a number of particles which correspond to the granular or molecular nature of materials.

The corresponding mathematical model results from associating to each point of the continuum a triad of vectors (directors). In the Cosserat theory each triad is assumed to rotate rigidly. As the deformation of the micro-medium is thus not considered, the displacement field of the macro-medium and the rotation field of the directors are enough to describe the deformation of the continuum.

In other more complex models, like Toupin's [27, 28] and Mindlin's [29] theories, the microstructure theory of elasticity of Green and Rivlin [30] and the general theory of microcontinua of Eringen and Suhubi [31] each director of the triad can stretch and rotate independently of the others. In other words, the micro-medium is allowed to deform independently of the macro-medium.

However we shall not be concerned with those theories. Cosserat theory alone will be described because of its connexions with the classical theories of shells and rods. We feel that its consideration is important if a general picture of different models of the theory of structures is to be drawn. On the other hand, the way will be open for the analysis of Cosserat continua by the finite element method.

5.2 Equilibrium Equations

In a granular material each grain is acted by external forces and moments and by the actions of other grains, which may also be reduced to forces and moments.

Passage to a continuum will be achieved if the discrete values of such forces and moments are replaced by continuous fields, according to the method of continuum mechanics.

It is understandable thus that, in a granular material, surface and body couples be considered in addition to tractions and body forces, and also that couple stresses, and not only ordinary stresses, be taken into consideration. The magnitude of such couples and couple-stresses depends on the size of the grains and can be neglected if the size is very small. We fall then into the field of application of classical continuum mechanics.

Let us associate to the body a system of curvilinear coordinates ξ_1, ξ_2, ξ_3, which, for sake of simplicity, are assumed to be orthogonal.

Denoting by $\vec{\sigma}_i$ and $\vec{\mu}_i$ the ordinary stress and couple-stress vectors acting on a positive face normal to a coordinate line i, the force and the moment equilibrium conditions for an infinitesimal tetrahedron with three faces perpendicular to coordinate lines and the fourth one to a unit vector $\vec{n}$ are

$$\vec{\sigma} = \sum_i \vec{\sigma}_i n_i \tag{5.2.1}$$

$$\vec{\mu} = \sum_i \vec{\mu}_i n_i \tag{5.2.2}$$

It is easy to conclude from (5.2.1) and (5.2.2) that the set of components σ_{ij} and μ_{ij} (generalized stresses) of vectors $\vec{\sigma}_i$ and $\vec{\mu}_i$ are components of two second order three-dimensional tensors: the stress tensor and the couple-stress tensor.

We remember now that the divergence theorem

for a domain V with boundary Σ can be expressed by

(5.2.3) $$\int_V \frac{1}{h} \sum_i \left(h_{\bar{i}} \vec{A}_i\right)_{,i} dV = \int_\Sigma \sum_i \left(A_i n_i\right) d\Sigma$$

in which $\vec{A}_1, \vec{A}_2, \vec{A}_3$, are vector functions admitting first derivatives with respect to ξ_1, ξ_2, ξ_3, and

(5.2.4) $$h = h_1 h_2 h_3$$

(5.2.5) $$h_{\bar{1}} = h_2 h_3$$

(5.2.6) $$h_{\bar{2}} = h_3 h_1$$

(5.2.7) $$h_{\bar{3}} = h_1 h_2$$

h_1, h_2, h_3, being the scale factors, i.e. functions of the coordinates such that

(5.2.8) $$\vec{r}_{,i} = h_i \vec{a}_i .$$

The equilibrium field equations can easily be established with the help of the divergence theorem, by considering the equilibrium conditions for an arbitrary portion V' of the body. Denoting by $\vec{f}$ and $\vec{g}$ the body force and couple density vectors, and by $\vec{\sigma}$ and $\vec{\mu}$ the traction and couple-stress vectors applied on the boundary Σ', the force and moment equilibrium conditions are respectively

(5.2.9) $$\int_{V'} \vec{f}\, dV + \int_{\Sigma'} \vec{\sigma}\, d\Sigma = \vec{0}$$

$$\int_V (\vec{g} + \vec{r} \wedge \vec{f})\, dV + \int_{\Sigma'} (\vec{\mu} + \vec{r} \wedge \vec{\sigma})\, d\Sigma = \vec{0} \qquad (5.2.10)$$

Introducing (5. 2. 1) and using the divergence theorem, equation (5. 2. 9) can be transformed into

$$\int_{V'} \left[\frac{1}{h} \sum_i (h_{\bar{i}}\, \vec{\sigma}_i)_{,i} + \vec{f} \right] dV = \vec{0} \qquad (5.2.11)$$

and, as (5.2.11) is true for any portion V' of V , we must have

$$\sum_i (h_{\bar{i}}\, \vec{\sigma}_i)_{,i} + h \vec{f} = \vec{0} \qquad (5.2.12)$$

Introducing now (5.2.1) and (5. 2. 2) in (5. 2. 10), and using the divergence theorem, we obtain

$$\int_{V'} \left\{ \vec{g} + \vec{r}_1 \vec{f} + \frac{1}{h} \sum_i \left[h_{\bar{i}} (\vec{\mu}_{,i} + \vec{r} \wedge \vec{\sigma}_i) \right]_{,i} \right\} dV' = \vec{0} \qquad (5.2.13)$$

and, as V' is an arbitrary portion of V , there follows

$$\vec{g} + \vec{r} \wedge \vec{f} + \frac{1}{h} \sum_i \left[\vec{h}_{\bar{i}} (\vec{\mu}_{,i} + \vec{r} \wedge \vec{\sigma}_i) \right]_{,i} = \vec{0} \qquad (5.2.14)$$

Eliminating $\vec{f}$ between (5. 2. 12) and (5.2.14), we obtain finally

$$\sum_i \left[(h_{\bar{i}}\, \vec{\mu}_i)_{,i} + h\, \vec{a}_i \wedge \vec{\sigma}_i \right] + h \vec{g} = \vec{0} \qquad (5.2.15)$$

Equations (5. 2. 12) and (5. 2. 14) are the equilibrium field equations. As

$$\sum_i \vec{a}_i \wedge \vec{\sigma}_i = (\sigma_{23} - \sigma_{32})\, \vec{a}_1 + (\sigma_{31} - \sigma_{13})\, \vec{a}_2 + (\sigma_{12} - \sigma_{21})\, \vec{a}_3 \qquad (5.2.16)$$

equation (5.2.15) shows that the stress tensor will be symmetric if the couple-stresses and the body density vanish in the body, as happens in the classical theory.

5.3 Strain-Displacement Equations

The strain-displacement equations can be established with the help of the principle of virtual work written for an arbitrary portion V' of V :

$$\int_{V'} \sum_i \left(\vec{\sigma}_i \cdot \vec{e}_i + \vec{\mu}_i \cdot \vec{k}_i\right) dV = \int_{V'} \left(\vec{f} \cdot \vec{u} + \vec{g} \cdot \vec{\Theta}\right) dV + \int_{\Sigma'} \left(\vec{\sigma} \cdot \vec{u} + \vec{\mu} \cdot \vec{\Theta}\right) d\Sigma$$

(5.3.1)

In equation (5.3.1) the body force and couple densities $\vec{f}$ and $\vec{g}$ and the surface force and couple densities $\vec{\sigma}$ and $\vec{\mu}$ are assumed to equilibrate the stress and couple-stress components σ_{ij} and μ_{ij} , respectively in V' and Σ'.

The components of $\vec{e}_i$ and $\vec{k}_i$, are the generalized strains e_{ij} and k_{ij} , which are assumed to be compatible with the generalized displacements u_i and Θ_i , i.e., the ordinary displacements and rotations.

We remark that the rotations which are considered among the generalized displacements are the rotations of the grains, i.e. the micro-rotations, and not the rotations of the macro-medium, i.e. the macro-rotations given by half the curl of the ordinary displacement vector.

Using (5.2.1) and (5.2.2), and the divergence theorem, the last integral in (5.3.1) can be transformed as follows

$$\int_{\Sigma'}(\vec{\sigma}\cdot\vec{u}+\mu\cdot\vec{\Theta})\,d\Sigma=\int_{\Sigma'}\sum_i(\vec{\sigma}_i\cdot\vec{u}+\vec{\mu}_i\cdot\vec{\Theta})n_i\,d\Sigma=\int_{V'}\frac{1}{h}\sum_i\left[h_{\bar{i}}(\vec{\sigma}_i\cdot\vec{u}+\vec{\mu}_i\cdot\vec{\Theta})\right]_{,i}dV.$$

(5.3.2)

Introducing in (5.3.1), and remarking that V' is arbitrary, there results

$$\sum_i(\vec{\sigma}_i\cdot\vec{e}_i+\vec{\mu}_i\cdot k_i)=\vec{f}\cdot\vec{u}+\vec{g}\cdot\vec{\Theta}+\frac{1}{h}\sum_i\left[h_{\bar{i}}(\vec{\sigma}_i\cdot\vec{u}+\vec{\mu}_i\cdot\vec{\Theta})\right]_{,i}.$$

(5.3.3)

Using the equilibrium equations (5.2.12) and (5.2.15), and remarking that

$$-\vec{a}_i\wedge\vec{\sigma}_i\cdot\vec{\Theta}=\vec{\sigma}_i\cdot\vec{a}_i\wedge\vec{\Theta} \qquad (5.3.4)$$

we obtain

$$\sum_i(\vec{\sigma}_i\cdot\vec{e}_i+\vec{\mu}_i\cdot\vec{k}_i)=\sum_i\left[\vec{\sigma}_i\cdot\left(\frac{\vec{u}_{,i}}{h_i}+\vec{a}_i\wedge\vec{\Theta}\right)+\vec{u}_i\cdot\frac{\vec{\Theta}_{,i}}{h_i}\right]. \qquad (5.3.5)$$

As (5.3.5) holds for any $\vec{\sigma}_i$ and $\vec{\mu}_i$, we must have

$$\vec{e}_i=\frac{\vec{u}_{,i}}{h_i}+\vec{a}_i\wedge\vec{\Theta} \qquad (5.3.6)$$

$$\vec{k}_i=\frac{\vec{\Theta}_{,i}}{h_i}. \qquad (5.3.7)$$

Equations (5.3.6) and (5.3.7) are the strain-displacement equations. We can also write

(5.3.8) $$e_{ij} = \frac{\vec{u}_{,i} \cdot \vec{a}_j}{h_i} - \vec{a}_i \wedge \vec{a}_j \cdot \vec{\Theta}$$

(5.3.9) $$k_{ij} = \frac{\vec{\Theta}_{,i} \cdot \vec{a}_j}{h_i}$$

Magnitudes e_{ij} and k_{ij} (generalized strains) are respectively called strains (or ordinary strains) and curvatures. It is possible to conclude from the invariance of $\sum_i (\vec{\sigma}_i \cdot \vec{e}_i)$ and $\sum_i (\vec{\mu}_i \cdot \vec{k}_i)$ that they are components of two tensors, the strain tensor and the curvature tensor.

5.4 Stress-Strain Equations

The stress-strain equations (constitutive equations) have the form

(5.4.1) $$\sigma_{ij} = \frac{\partial W_v}{\partial e_{ij}}$$

(5.4.2) $$\mu_{ij} = \frac{\partial W_v}{\partial k_{ij}}$$

in which W_v is the strain energy density in V .

If the material is isotropic and if the initial stresses are assumed to vanish, W_v must be a function of the fundamental invariants of the generalized strain tensors. On the other hand, if the material is assumed to be linearly behaved,

W_V must be a quadratic function of the strains and thus have the form

$$W_V = \frac{1}{2}(\lambda+2\mu)e^2 - 2\mu e'_s - 2\rho e'_a + \frac{1}{2}(\varphi+2\psi)k^2 - $$
$$- 2\psi k'_s - 2\chi k'_a + \pi e k \qquad (5.4.3)$$

in which e is the first invariant of the strain tensor

$$e = e_{11} + e_{22} + e_{33}, \qquad (5.4.4)$$

e'_s and e'_a are the second invariants of the symmetric and skew-symmetric parts of the strain tensor

$$e'_s = e_{22}e_{33} + e_{33}e_{11} + e_{11}e_{22} - \frac{1}{4}\left[(e_{23}+e_{32})^2 + (e_{31}+e_{13})^2 + (e_{12}+e_{21})^2\right] \qquad (5.4.5)$$

$$e'_a = -\frac{1}{4}\left[(e_{23}-e_{32})^2 + (e_{31}-e_{13})^2 + (e_{12}-e_{21})^2\right] \qquad (5.4.6)$$

and k, k'_s and k'_a are the corresponding invariants of the curvature tensor.

Introducing (5.4.3) in (5.4.1) and (5.4.2), we obtain

$$\sigma_{ij} = 2\mu\,\frac{e_{ij}+e_{ji}}{2} + 2\rho\,\frac{e_{ij}-e_{ji}}{2} + (\lambda e + \pi k)\,\delta_{ij} \qquad (5.4.7)$$

$$\mu_{ij} = 2\psi\,\frac{k_{ij}+k_{ji}}{2} + 2\chi\,\frac{k_{ij}-k_{ji}}{2} + (\varphi k + \pi e)\,\delta_{ij} \qquad (5.4.8)$$

It is now reasonable to assume [32] that the couple-stresses vanish if the body is submitted to a uniform field of strain characterized by

(5.4.9) $\qquad e_{11} = e_{22} = e_{33} =$ an arbitrary constant

all the remaining generalized strains being equal to zero. This assumption leads to $\pi = 0$, and thus to six elastic constants. [33] The stress-strain equations become

$$\sigma_{ij} = 2\mu \frac{e_{ij} + e_{ji}}{2} + 2\rho \frac{e_{ij} - e_{ji}}{2} + \lambda e \delta_{ij} \tag{5.4.10}$$

$$\mu_{ij} = 2\psi \frac{k_{ij} + k_{ji}}{2} + 2\chi \frac{k_{ij} - k_{ji}}{2} + \varphi k \delta_{ij} . \tag{5.4.11}$$

The strain-stress equations are

$$e_{ij} = \frac{1}{2\mu} \frac{\sigma_{ij} + \sigma_{ji}}{2} + \frac{1}{2\rho} \frac{\sigma_{ij} - \sigma_{ji}}{2} - \frac{\lambda}{2\mu(2\mu + 3\lambda)} s \delta_{ij}$$

(5.4.12)

$$k_{ij} = \frac{1}{2\psi} \frac{\mu_{ij} + \mu_{ji}}{2} + \frac{1}{2\chi} \frac{\mu_{ij} - \mu_{ji}}{2} - \frac{\varphi}{2\psi(2\psi + 3\varphi)} m \delta_{ij}$$

(5.4.13)

in which s and m are the first invariants of the stress and couple-stress tensors.

The strain energy density can be given the form

$$2W_v = \sum_{i,j} (\sigma_{ij} e_{ij} + \mu_{ij} k_{ij}) = \sum_{i,j} \Big(2\mu \frac{e_{ij} + e_{ji}}{2} + 2\rho \frac{e_{ij} - e_{ji}}{2} +$$

(5.4.14)

$$+\lambda e\delta_{ij}\Big)e_{ij}+\Big(2\psi\,\frac{k_{ij}+k_{ji}}{2}+2\chi\,\frac{k_{ij}-k_{ji}}{2}+\varphi k\delta_{ij}\Big)k_{ij}\,. \tag{5.4.14}$$

But, as

$$\Big(2\mu\,\frac{e_{ij}+e_{ji}}{2}+2\rho\,\frac{e_{ij}-e_{ji}}{2}+\lambda e\delta_{ij}\Big)e_{ij}= \\ =(\mu+\rho)e_{ij}e_{ij}+(\mu-\rho)e_{ij}e_{ji}+\lambda e^2 \tag{5.4.15}$$

and

$$\sum_{i,j} e_{ij}e_{ij}=\sum_{i,j} e_{ji}e_{ji} \tag{5.4.16}$$

(5.4.14) may be transformed into

$$2W_v=\frac{1}{2}\mu\sum_{i,j}(e_{ij}+e_{ji})^2+\frac{1}{2}\rho\sum_{i,j}(e_{ij}-e_{ji})^2+\lambda e^2+\frac{1}{2}\psi\sum_{i,j}(k_{ij}+k_{ji})^2+ \\ +\frac{1}{2}\chi\sum_{i,j}(k_{ij}-k_{ji})^2+\varphi k^2 \tag{5.4.17}$$

As, for sake of stability, W_V must be a positive definitive function of the strains, it is easy to conclude from (5.4.17) that the elastic constants $\mu, \rho, \lambda, \psi, \chi$ and φ must also be positive.

5.5 Balance of Equations and Unknowns in the General Theory

In the three-dimensional Cosserat theory, we have six equilibrium equations (5.2.12) and (5.2.15), eighteen strain displacement equations, (5.3.6) and (5.3.7) and eighteen stress-strain equations (5.4.10) and (5.4.11). The total number of equations is thus forty-two.

The unknowns are also 42: eighteen generalized stresses σ_{ij} and μ_{ij}, eighteen generalized strains e_{ij} and k_{ij} and six generalized displacements u_i and θ_i.

In what regards the boundary conditions, we have six for each boundary point. Denoting by Σ_1 and Σ_2 the portions of the boundary where stresses and displacements are respectively prescribed, the boundary conditions are

$$\left.\begin{aligned} \sigma_i &= \bar{\sigma}_i \\ \mu_i &= \bar{\mu}_i \end{aligned}\right\} \text{ on } \Sigma_1 \tag{5.5.1}$$

$$\left.\begin{aligned} u_i &= \bar{u}_i \\ \theta_i &= \bar{\theta}_i \end{aligned}\right\} \text{ on } \Sigma_2 \tag{5.5.2}$$

5.6 Simplified Theory

A usual simplification [25] consists in assuming the micro-rotations equal to the macro-rotations, i.e.

$$\vec{\theta} = \frac{1}{2} \operatorname{curl} \vec{u} . \tag{5.6.1}$$

The expression for curl $\vec{u}$ is [34]

$$\operatorname{curl} \vec{u} = \frac{1}{h} \begin{vmatrix} h_1 \vec{a}_1 & h_2 \vec{a}_2 & h_3 \vec{a}_3 \\ \dfrac{\partial}{\partial \xi_1} & \dfrac{\partial}{\partial \xi_2} & \dfrac{\partial}{\partial \xi_3} \\ h_1 u_1 & h_2 u_2 & h_3 u_3 \end{vmatrix} = \frac{1}{u} \Big\{ h_1 \vec{a}_1 \big[(h_3 u_3)_{,2} - \tag{5.6.2}$$

$$- (h_2 u_2)_{,3}\Big] + h_2 \vec{a}_2 \Big[(h_1 u_1)_{,3} - (h_3 u_3)_{,1}\Big] + h_3 \vec{a}_3 \Big[(h_2 u_2)_{,1} - (h_1 u_1)_{,2}\Big]\Big\}$$

(5.6.2)

As it is possible to demonstrate that

$$\frac{1}{h_i h_j}\Big[(h_i u_i)_{,j} - (h_j u_j)_{,i}\Big] = \frac{\vec{u}_{,j} \cdot \vec{a}_i}{h_j} - \frac{\vec{u}_{,i} \cdot \vec{a}_j}{h_i} \qquad (5.6.3)$$

(5.6.1) can be written

$$2\Theta = \left(\frac{\vec{u}_{,2} \cdot \vec{a}_3}{h_2} - \frac{\vec{u}_{,3} \cdot \vec{a}_2}{h_3}\right)\vec{a}_1 + \left(\frac{\vec{u}_{,3} \cdot \vec{a}_1}{h_3} - \frac{\vec{u}_{,1} \cdot \vec{a}_3}{h_1}\right)\vec{a}_2 + \left(\frac{\vec{u}_{,1} \cdot \vec{a}_2}{h_1} - \frac{\vec{u}_{,2} \cdot \vec{a}_1}{h_2}\right)\vec{a}_3$$

(5.6.4)

This means that

$$\vec{a}_i \wedge \vec{a}_j \cdot \vec{\Theta} = \frac{1}{2}\left(\frac{\vec{u}_{,i} \cdot \vec{a}_j}{h_i} - \frac{\vec{u}_{,j} \cdot \vec{a}_i}{h_j}\right) \qquad (5.6.5)$$

Introducing in (5.3.8), we obtain

$$e_{ij} = \frac{1}{2}\left(\frac{\vec{u}_{,i} \cdot \vec{a}_j}{h_i} + \frac{\vec{u}_{,j} \cdot \vec{a}_i}{h_j}\right) = e_{ji} \, . \qquad (5.6.6)$$

This means that the simplifying assumption (5.6.1) implies the symmetry of the strain tensor.

On the other hand, the curvatures become functions of the displacements. Indeed, introducing (5.6.1) in (5.3.9), there results

$$\vec{k}_i = \frac{1}{2} \frac{(\mathrm{curl}\, \vec{u})_{,i}}{h_i} \qquad (5.6.7)$$

The simplified theory gives no information about the isotropic parts of the couple-stress and curvature tensors. Indeed (5.6.7) implies

$$k = k_{11} + k_{22} + k_{33} = \sum_i \vec{k}_i \cdot \vec{a}_i = \frac{1}{2} \sum_i \frac{(\text{curl } \vec{u})_{,i} \cdot \vec{a}_i}{h_i} = 0 \tag{5.6.8}$$

and thus, by virtue of (5.4.11),

$$m = \mu_{11} + \mu_{22} + \mu_{33} = (2\psi + 3\varphi)k = 0 \tag{5.6.9}$$

For this reason, the simplified theory is usually called the indeterminate couple-stress theory [25] .

The assumption (5.6.1) which is in the base of the simplified theory would be exact if the elastic constant ρ was infinitely large. Then indeed e_{ij} would be really equal to e_{ji} and this would imply (5.6.1).

5.7 Balance of Equations in the Simplified Theory

In the simplified theory, the three supplementary equations (5.6.1) are introduced which generally are not compatible with the remaining ones. This means that three of such remaining equations must be disregarded.

In order that neither equilibrium nor compatibility be violated the equations to be disregarded must be stress-strain equations. We replace then the nine independent stress-strain equations (5.6.10) by the six independent equations

$$\frac{\sigma_{ij}+\sigma_{ji}}{2} = 2\mu\, e_{ij} + \lambda\, e\, \delta_{ij} \tag{5.7.1}$$

These equations are obtained by averaging the equation corresponding to σ_{ij} and the equation corresponding to σ_{ji} .

We remark that, if equations (5.4.10) were not replaced by equations (5.7.1), the symmetry of the strain tensor would imply the symmetry of the stress tensor, which would violate the equilibrium equation (5.2.15).

As equations (5.4.11) remain intact, the total number of stress-strain independent equation becomes 15 in the simplified theory.

In what concerns the strain-displacement equations only six of the equations (5.3.6) are independent, because of the e_{ij} being symmetric. We have thus also 15 independent equations.

The independent magnitudes which are involved in all these equations are 3 displacements u_1, u_2, u_3, six ordinary strains e_{ij} , 9 curvatures k_{ij} , 6 components of the symmetric part of the stress tensor $\frac{\sigma_{ij}+\sigma_{ji}}{2}$ and 9 components of the couple-stress tensor.

We have thus, up to this moment, 30 equations and 33 unknowns. The missing 3 equations must be equilibrium equations involving the components of the symmetric part of the stress tensor, $\sigma_{ij}^{s} = \dfrac{\sigma_{ij}+\sigma_{ji}}{2}$, and the couple-stresses μ_{ij} . Such equations can be found by eliminating the skew-symmetric part of the stress tensor between (5.2.12) and (5.2.15) as follows.

Equation (5.2.12) can be transformed into

$$\sum_i \left(k_{\bar{i}}\, \vec{\sigma}_i^{s}\right)_{,i} + \sum_i \left(h_{\bar{i}}\, \vec{\sigma}_i^{a}\right)_{,i} + h\vec{F} = \vec{0} \tag{5.7.2}$$

in which vectors $\vec{\sigma}_i^{s}$ and $\vec{\sigma}_i^{a}$ correspond respectively to the symmetric and skew-symmetric parts of the stress tensor.

As

$$\sum_j (\vec{a}_j \wedge \vec{\sigma}_j^{a}) \wedge \vec{a}_i = \sum_j \left[(\vec{a}_i \cdot \vec{a}_j)\,\vec{\sigma}_j^{a} - (\vec{a}_i \cdot \vec{\sigma}_j^{a})\,\vec{a}_j\right] - \sum_j \left(\delta_{ij}\vec{\sigma}_j^{a} - \sigma_{ji}^{a}\vec{a}_j\right) =$$

$$= \vec{\sigma}_i^{a} + \sum_j \sigma_{ij}^{a}\, \vec{a}_j = \vec{\sigma}_i^{a} + \vec{\sigma}_i^{a} = 2\vec{\sigma}_i^{a} \tag{5.7.3}$$

equation (5.7.2) can be transformed into

$$\sum_i \left(h_{\bar{i}}\, \vec{\sigma}_i^{s}\right)_{,i} + \frac{1}{2} \sum_i \left[h_{\bar{i}} \left(\sum_j \vec{a}_j \wedge \vec{\sigma}_j^{a}\right) \wedge \vec{a}_i\right]_{,i} + h\vec{F} = \vec{0} \tag{5.7.4}$$

On the other hand, (5.2.15) yields

$$\sum_j (\vec{a}_j \wedge \vec{\sigma}_j^{a}) = \sum_j \vec{a}_j \wedge \vec{\sigma}_j = -\frac{1}{h} \sum_j \left(h_{\bar{j}}\, \vec{\mu}_j\right)_{,j} - \vec{g} \tag{5.7.5}$$

Introducing (5.7.5) in (5.7.4), we obtain

$$\sum_i \left(h_{\bar{i}}\, \vec{\sigma}_i^{s}\right)_{,i} + \frac{1}{2} \sum_{ij} \left[\frac{1}{h_i}\, \vec{a}_i \wedge \left(h_{\bar{j}}\, \vec{\mu}_j\right)_{,j}\right]_{,i} + h\vec{F} + \frac{1}{2} \sum_i \left(h_{\bar{i}}\, \vec{a}_i \wedge \vec{g}\right)_{,i} = \vec{0}$$

(5.7.6)

which represents the three equilibrium equations needed.

5.8 Boundary Conditions in the Simplified Theory

The displacement boundary conditions of the simplified theory are simply

$$\vec{u} = \vec{\bar{u}} \tag{5.8.1}$$

$$\vec{\Theta} \wedge \vec{n} = \vec{\bar{r}} \tag{5.8.2}$$

because, by virtue of (5.6.1), the component of rotation normal to the boundary can be expressed in terms of the displacements specified on the boundary. $\vec{\bar{u}}$ and $\vec{\bar{r}}$ are vectors prescribed on Σ_2 .

The stress boundary conditions can be determined by considering the work $\delta\tau$ done by the tractions and couple-stresses applied on the boundary when the displacement and rotation fields are given virtual variations $\delta\vec{u}$ and $\delta\vec{\Theta}$. Considering that $\delta\vec{u} = 0$ on Σ_2 , we have

$$\delta\tau = \int_{\Sigma} \left(\vec{\sigma}\cdot\delta\vec{u} + \vec{\mu}\cdot\delta\vec{\Theta}\right) d\Sigma = \int_{\Sigma_1} \left(\vec{\sigma}\cdot\delta\vec{u} + \delta\vec{\Theta}\right) d\Sigma \tag{5.8.3}$$

Let now $\delta\Theta_n$ and μ_n denote the components of $\delta\vec{\Theta}$ and $\vec{\mu}$ in the direction of $\vec{n}$. We have

$$\vec{\mu}\cdot\delta\vec{\Theta} = \mu_n \delta\Theta_n + \vec{\mu}\wedge\vec{n}\cdot\delta\vec{\Theta}\wedge\vec{n} \tag{5.8.4}$$

By virtue of (5.6.1)

$$\delta\Theta_n = \frac{1}{2}\,\text{curl}\,\delta\vec{u}\cdot\vec{n} \tag{5.8.5}$$

and thus

$$\delta\Theta_n\mu_n = \frac{1}{2}\,\text{curl}\,\vec{u}\cdot\mu_n\vec{n} = \frac{1}{2}\,\text{curl}\left(\mu_n\vec{n}\right)\cdot\delta\vec{u} + \frac{1}{2}\,\text{div}\left(\mu_n\vec{n}\wedge\delta\vec{u}\right) \tag{5.8.6}$$

But

$$\mu_n\vec{n} = \text{curl}\left(\mu_n h_n\frac{\vec{n}}{h_n}\right) = \mu_n h_n\,\text{curl}\,\frac{\vec{n}}{h_n} + \text{grad}\left(\mu_n h_n\right)\wedge\frac{\vec{n}}{h_n} =$$

$$= \text{grad}\left(\mu_n h_n\right)\wedge\frac{\vec{n}}{h_n} = \frac{1}{h_n}\,\text{grad}_\Sigma\left(\mu_n h_n\right)\wedge\vec{n} \tag{5.8.7}$$

in which grad_Σ denotes the gradient on the two-dimensional domain Σ.

On the other hand, the divergence theorem for domain Σ_1 permits to write

$$\int_{\Sigma_1}\text{div}\left(\mu_n\vec{n}\wedge\delta\vec{u}\right)d\Sigma = \int_{\Gamma_1}\mu_n\vec{n}\wedge\delta\vec{u}\cdot\vec{\nu}\,d\Gamma \tag{5.8.8}$$

in which Γ_1 denotes the contour of Σ_1 and $\vec{\nu}$ the unit vector normal to such contour and tangent to Σ at a given point. As $\delta\vec{u}$ is equal to zero on Σ_2 and thus on Γ_1, it follows that

$$\int_{\Sigma_1}\text{div}\left(\mu_n\vec{n}\wedge\delta\vec{u}\right)d\Sigma = 0 \tag{5.8.9}$$

The case in which only stresses are prescribed on the boundary can be considered as the limit case of a process in which Σ_2 decreases continuously and thus Γ_1 shrinks to a single point. The second member (5.8.8) then becomes smaller and vanishes in the limit case, to that (5.8.9) is still valid.

By virtue of (5.8.6), (5.8.7) and (5.8.9),

$$\int_{\Sigma_1} \delta\Theta_n \mu_n \, d\Sigma = \int_{\Sigma_1} \frac{1}{2h_n} \operatorname{grad}_\Sigma (\mu_n h_n) \wedge \vec{n} \cdot \delta\vec{u} \, d\Sigma \tag{5.8.10}$$

and thus

$$\delta\tau = \int_{\Sigma_1} \left\{ \left[\vec{\sigma} + \frac{1}{2h_n} \operatorname{grad}_\Sigma (\mu_n h_n) \wedge \vec{n} \right] \delta\vec{u} + \vec{\mu} \wedge \vec{n} \cdot \delta\vec{\Theta} \wedge \vec{n} \right\} d\Sigma \tag{5.8.11}$$

This means that the stress boundary conditions are

$$\vec{\sigma} + \frac{1}{2h_n} \operatorname{grad}_\Sigma (\mu_n h_n) \wedge \vec{n} = \vec{s} \tag{5.8.12}$$

$$\vec{\mu} \wedge \vec{n} = \vec{t} \tag{5.8.13}$$

in which $\vec{s}$ and $\vec{t}$ are vectors prescribed on Σ_1.

5.9 The Classical Theory of Elasticity

The classical theory results from the Cosserat theory if ψ, χ and φ are assumed to vanish. Then indeed the couple-stresses vanish and the equilibrium equation (5.2.15) becomes

(5.9.1) $$\sum_i h_{\bar{i}} \vec{a}_i \wedge \vec{\sigma}_i + h\vec{g} = \vec{0}$$

which implies

(5.9.2) $$\sigma_{ij} = \sigma_{ji}$$

if $\vec{g}$ is also assumed to vanish.

(5.9.2) implies

(5.9.3) $$e_{ij} = e_{ji}$$

by virtue of (5.4.12).

The stress and strain tensors are thus both symmetric.

By virtue of (5.3.8) and (5.9.3), the strain displacement equations have the form (5.6.6) and the rotations can be expressed in terms of the displacements by (5.6.1) or (5.6.5).

Let now [35]

(5.9.4) $$\ell = \sqrt{\frac{\psi}{\mu}}$$

Magnitude ℓ, which has the dimensions of a length, is usually called the characteristic length. In a granular material ℓ is connected with the size of the grains.

Assuming χ and φ to have the same order of magnitude as ψ, all the three constants will tend to zero with ℓ. Classical elasticity will be valid thus if the characteristic length is small in comparison with the typical dimensions of the body or with the wave-lengths in wave propagation problems.

6. TWO-DIMENSIONAL MODEL

6.1 Introduction

A general two-dimensional theory will be developed in the same lines as the three-dimensional one considered before. The two-dimensional domain will be denoted by S and its boundary by Γ .

Curvilinear coordinates ξ_1, ξ_2, ξ_3, will be considered such that:

a) the domain S is contained in the coordinate surface $\xi_3 = 0$;

b) the coordinate lines on the surface $\xi_3 = 0$ are principal lines of the surface;

c) the coordinate lines corresponding to constant values of ξ_1 and ξ_2 are straight lines normal to the surface $\xi_3 = 0$;

d) the scale factor h_3 is constant in the whole space. h_3 will be assumed equal to unity for sake of simplicity.

It is possible to demonstrate that a system of coordinates obeying such conditions is orthogonal in the whole space.

The unit vectors tangent to the coordinate lines do not depend on ξ_3 , i.e.

(6.1.1) $$\vec{a}_{i,3} = \vec{0}$$

and, if we denote by h_1^0, h_2^0, R_1^0 and R_2^0 the scale factors and radii of curvature of points on the middle surface (°), we have

$$\vec{a}_{3\alpha} = \frac{h_\alpha^0}{R_\alpha^0} \vec{a}_\alpha \qquad (6.1.2)$$

In equation (6.1.2) and along the whole chapter, greek indices are assumed to take up only the values 1 or 2, while latin indices can take the values 1, 2 or 3.

Let

$$\lambda_\alpha = \frac{h_\alpha}{h_\alpha^0} \qquad (6.1.3)$$

denote the ratio between a scale factor $h_\alpha(\xi_1, \xi_2, \xi_3)$ at an arbitrary point and the scale factor $h_\alpha(\xi_1\ \xi_2)$ at the point of S with the same coordinates.

It is possible to prove [18] that, under the assumptions which have been made,

$$\lambda_\alpha = 1 + \frac{\xi_3}{R_\alpha^0} \qquad (6.1.4)$$

the theory will refer to bodies called shells.

The boundary Σ of a shell, considered as a three-dimensional body is composed of two parts. The first

---

(°) Superscript ° will be used to denote magnitudes referring to S. As $\xi_3 = 0$ on S, such magnitudes can be considered as functions of ξ_1 and ξ_2 alone. The unit vectors $\vec{a}_\alpha$ tangent to S do not need any superscript on account of (6.1.1).

part, Σ_f , is formed by two surfaces with equations $\xi_3 = \pm\frac{1}{2}t(\xi_1,\xi_2)$. Magnitude t is called the thickness of the shell. For sake of simplicity, t will be assumed constant.

The second part, Σ_b , is formed by one or more surfaces with analytical expressions of the type $f(\xi_1,\xi_2)=0$.

The boundary Γ of S is nothing else than the intersection of Σ_b with the surface $\xi_3=0$. Each point of Σ_b can be referred to a system of coordinates, ξ and ξ_3 . As $h_3=1$, the scale factor h_Γ corresponding to ξ is fixed everywhere on Σ_b, once h_Γ^0 , i.e. the scale factor along Γ , is fixed. We shall take $h_\Gamma^0=1$.

Let λ be defined by

$$\lambda = \frac{h_\Gamma}{h_\Gamma^0} \tag{6.1.5}$$

λ plays thus the same role on Σ_b as magnitudes λ_α on the coordinate surfaces.

Denote by $\vec{n}$ the unit vector normal to Σ . $\vec{n}$ is normal to $\vec{a}_3$ on Σ_b. This means that, for points on Σ_b, $n_3=0$.

Let $\vec{n}^0$ be the unit vector to Σ_b on Γ . $\vec{n}^0$ is normal to $\vec{a}_3$ and thus tangent to S . It can be shown [18] that

$$\lambda n_\alpha = \lambda_{\bar{\alpha}} n_\alpha^0 \tag{6.1.6}$$

in which $\bar{\alpha}$ is equal to 1 if α is equal to 2 and equal to 2 if α

is equal to 1.

6.2 Equilibrium Equations

The equilibrium equations for the two-dimensional case can be established either directly, using the divergence theorem for the two-dimensional domains

$$\int_S \frac{1}{h^0_\alpha} \sum_\alpha \left(h^0_{\bar{\alpha}} \vec{A}_\alpha \right)_{,\alpha} dS = \int_\Gamma \sum_\alpha \vec{A}_\alpha n^0_\alpha \, d\Gamma \qquad (6.2.1)$$

or by integrating the three-dimensional equations (5.2.12) and (5.2.15) in the thickness of the shell.

The first method will be used in order that the analogy with the three-dimensional case be more striking. The second method will be exemplified however in the deduction of the boundary equilibrium equations.

Let us consider equations (5.2.1) and (5.2.2) for points located on Σ_b. Multiplying them by λ and integrating, we obtain, by virtue of (6.1.6)

$$\int_{-t/2}^{t/2} \lambda \vec{\sigma} \, d\xi_3 = \int_{-t/2}^{t/2} \lambda \sum_i \vec{\sigma}_i n_i \, d\xi_3 = \int_{-t/2}^{t/2} \lambda \sum_\alpha \vec{\sigma}_\alpha n_\alpha \, d\xi_3 = \sum_\alpha \int_{-t/2}^{t/2} \lambda_{\bar{\alpha}} \vec{\sigma}_\alpha \, d\xi_3 \, n^0_\alpha \qquad (6.2.2)$$

Making

$$\vec{N} = \int_{-t/2}^{t/2} \lambda \vec{\sigma} \, d\xi_3 \qquad (6.2.3)$$

(6.2.4)
$$\vec{N}_\alpha = \int_{t/2}^{t/2} \lambda_{\bar{\alpha}} \, \vec{\sigma}_\alpha \, d\xi_3$$

equation (6.2.3) can be transformed into

(6.2.5)
$$\vec{N} = \sum_\alpha \vec{N}_\alpha n_\alpha^0$$

The equation

(6.2.6)
$$\vec{M} = \sum_\alpha \vec{M}_\alpha n_\alpha^0$$

in which

(6.2.7)
$$\vec{M} = \int_{-t/2}^{t/2} \lambda \left(\vec{a}_3 \, \xi_3 \wedge \vec{\sigma} + \vec{\mu} \right) d\xi_3$$

(6.2.8)
$$\vec{M}_\alpha = \int_{-t/2}^{t/2} \lambda_{\bar{\alpha}} \left(\vec{a}_3 \, \xi_3 \wedge \vec{\sigma}_\alpha + \vec{\mu}_\alpha \right) d\xi_3$$

can be obtained also by integrating a linear combination of equations (5.2.1) and (5.2.2).

It is easy to conclude from (6.2.5) and (6.2.6) that the components $N_{\alpha\beta}$ and $M_{\alpha\beta}$ of vectors $\vec{N}_\alpha$ and $\vec{M}_\alpha$ are also components of two second-order two-dimensional tensors. $N_{\alpha 3}$ and $M_{\alpha 3}$ are components of two vectors.

Let us illustrate now the second method in the deduction of the equilibrium field equations.

Let S' be the portion of S corresponding to a fragment of the shell. Let Γ' be the boundary of S' .

The stresses and couple-stresses in the boundary of the fragment are statically equivalent to forces and moments distributed on Γ with densities $\vec{N}$ and $\vec{M}$ given by equations (6.2.6) (fig. 6.2.1).

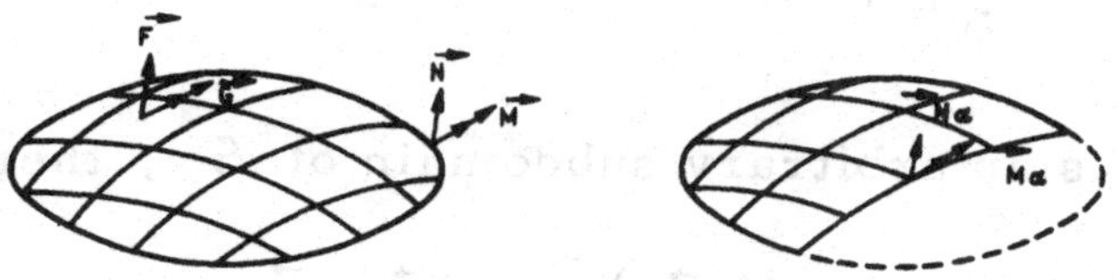

Fig. 6.2.1

Body forces and couples distributed in the volume of shell and on the faces are also statically equivalent to forces and moments, distributed on S, with densities $\vec{F}$ and $\vec{G}$ given by

$$\vec{F} = \int_{-t/2}^{t/2} \lambda_1 \lambda_2 \vec{f} \, d\xi_3 \tag{6.2.9}$$

$$\vec{G} = \int_{-t/2}^{t/2} \lambda_1 \lambda_2 \left(\vec{a}_3 \xi_3 \wedge \vec{f} + \vec{g} \right) d\xi_3 . \tag{6.2.10}$$

The force and moment equilibrium conditions for the shell fragment are thus

$$\int_{S'} \vec{F} \, dS + \int_{\Gamma'} \vec{N} \, d\Gamma = \vec{0} \tag{6.2.11}$$

$$\int_{S'} \left(\vec{G} + \vec{r} \wedge \vec{F} \right) dS + \int_{\Gamma'} \left(\vec{M} + \vec{r} \wedge \vec{N} \right) d\Gamma = 0 \tag{6.2.12}$$

Using (6.2.5) and the divergence theorem (6.2.1), (6.2.11) can easily be transformed into

$$\int_{S'} \left[\frac{1}{h^0} \sum_\alpha \left(h^0_{\bar{\alpha}} \vec{N}_\alpha \right)_{,\alpha} + \vec{F} \right] dS = \vec{0} \tag{6.2.13}$$

and, as S' is an arbitrary subdomain of S, there results

$$\sum_\alpha \left(h^0_{\bar{\alpha}} \vec{N}_\alpha \right)_{,\alpha} + h^0 \vec{F} = \vec{0} \tag{6.2.14}$$

Using (6.2.6) and the divergence theorem, we obtain easily, from (6.2.12), the second equilibrium equation

$$\sum_\alpha \left[\left(h^0_{\bar{\alpha}} \vec{M}_\alpha \right)_{,\alpha} + h^0 \vec{a}^0_\alpha \wedge \vec{N}_\alpha \right] + h^0 \vec{G} = \vec{0} \tag{6.2.15}$$

Equations (6.2.14) and (6.2.15) involve 12 generalized stresses of the two-dimensional general theory $N_{\alpha\beta}$, $N_{\alpha 3}$, $M_{\alpha\beta}$ and $M_{\alpha\beta}$.

6.3 Strain-Displacement Equations

The strain displacement equations can easily be deduced from the principle of virtual work written for the two-dimensional case

$$\int_{S'} \sum_\alpha \left(\vec{N}_\alpha \cdot \vec{E}_\alpha + \vec{M}_\alpha \cdot \vec{K}_\alpha \right) dS = \int_{S'} \left(\vec{F} \cdot \vec{U} + \vec{G} \cdot \vec{\Theta} \right) dS + \int_{\Gamma'} \left(\vec{N} \cdot \vec{U} + \vec{M} \cdot \vec{\Theta} \right) d\Gamma$$

(6.3.1)

The 12 generalized strains $E_{\alpha i}$ and $K_{\alpha i}$ and the 6 generalized displacements U_i and Θ_i are respectively defined as multipliers of the generalized stresses and forces in the expression of work.

Introducing (6.2.5) and (6.2.6) and using the divergence theorem together with equations (6.2.14) and (6.2.15) we obtain

$$\sum_\alpha \left(\vec{N}_\alpha \cdot \vec{E}_\alpha + \vec{M}_\alpha \cdot \vec{K}_\alpha\right) = \sum_\alpha \left[\vec{N}_\alpha \cdot \left(\frac{\vec{U}_{,\alpha}}{h^0_\alpha} + \vec{a}_\alpha \wedge \vec{\Theta}\right) + \vec{M}_\alpha \cdot \frac{\vec{\Theta}_{,\alpha}}{h^0_\alpha}\right] \qquad (6.3.2)$$

As (6.3.2) holds for any $\vec{N}_\alpha$ and $\vec{M}_\alpha$ we have

$$\vec{E}_\alpha = \frac{\vec{U}_{,\alpha}}{h^0_\alpha} + \vec{a}_\alpha \wedge \vec{\Theta} \qquad (6.3.3)$$

$$\vec{K}_\alpha = \frac{\vec{\Theta}_{,\alpha}}{h^0_\alpha} \qquad (6.3.4)$$

which are the strain-displacement equations.

We can also write

$$E_{\alpha i} = \frac{\vec{U}_{,\alpha} \cdot \vec{a}_i}{h^0_\alpha} - \vec{a}_\alpha \wedge \vec{a}_i \cdot \vec{\Theta} \qquad (6.3.5)$$

$$K_{\alpha i} = \frac{\vec{\Theta}_{,\alpha} \cdot \vec{a}_i}{h^0_\alpha} \, . \qquad (6.3.6)$$

Magnitudes $E_{\alpha\beta}$ and $K_{\alpha\beta}$ are components of two second-order two-dimensional tensors. Magnitudes $E_{\alpha 3}$ and $K_{\alpha 3}$ are components of two two-dimensional vectors.

6.4 Stress-Strain Equations

Let W_S be the strain energy density on S. If the material is isotropic and and the initial strains are assumed to vanish, then W_S can be expressed in terms of the fundamental invariants of the two tensors with components $E_{\alpha\beta}$ and $K_{\alpha\beta}$ and of the two vectors with components $E_{\alpha 3}$ and $K_{\alpha 3}$ Thus, in the linear case, W_S will have the form

$$(6.4.1)\qquad \begin{aligned} W_S = {} & \frac{1}{2}(\bar{\lambda}+\bar{\mu}_e)E^2 - 2\bar{\mu}_e E'_s - 2\bar{\rho}E'_a + \bar{\nu}E_3 + \\ & + \frac{1}{2}(\bar{\varphi}+2\bar{\psi})K^2 - 2\bar{\psi}K'_s - 2\bar{\chi}K'_a + \bar{\tau}K_3 \end{aligned}$$

in which

$$(6.4.2)\qquad E = E_{11} + E_{22}$$

$$(6.4.3)\qquad E'_s = E_{11}E_{22} - \frac{1}{4}(E_{12}+E_{21})^2$$

$$(6.4.4)\qquad E'_a = -\frac{1}{4}(E_{12}-E_{21})^2$$

$$(6.4.5)\qquad E_3 = E_{13}^2 + E_{23}^2$$

and K, K'_s, K'_a and K_3 have analogous definitions.

No term appears in (6.4.1) proportional to KE for a reason similar to the one which was given in 5.4. No moments are indeed necessary for introducing a constant generalized strain field characterized by arbitrary values of E_{11} and E_{22}, all the remaining generalized strains being equal to zero. The total number of constants is thus 8.

The stress-strain relations corresponding to (6.4.1) can be determined by differentiating W_s in order to the generalized strains.

There results

$$N_{\alpha\beta} = 2\bar{\mu}\,\frac{E_{\alpha\beta}+E_{\beta\alpha}}{2} + 2\bar{\rho}\,\frac{E_{\alpha\beta}-E_{\beta\alpha}}{2} + \bar{\lambda} E \delta_{\alpha\beta} \qquad (6.4.6)$$

$$N_{\alpha\beta} = 2\bar{\nu} E_{\alpha 3} \qquad (6.4.7)$$

$$M_{\alpha\beta} = 2\bar{\psi}\,\frac{K_{\alpha\beta}+K_{\beta\alpha}}{2} + 2\bar{\chi}\,\frac{K_{\alpha\beta}-K_{\beta\alpha}}{2} + \bar{\varphi} K \delta_{\alpha\beta} \qquad (6.4.8)$$

$$M_{\alpha 3} = 2\bar{\tau} K_{\alpha 3} \qquad (6.4.9)$$

It can be proved, as we did in 5.4 that, for sake of stability, all the elastic constants must be positive.

6.5 Balance of Equations and Unknowns in the General Theory

In the general two-dimensional theory, we have 6 equilibrium equations, (6.2.14) and (6.2.15), 12 strain-displacement equations (6.3.3) and (6.3.4), and 12 stress-strain equations, (6.4.6), (6.4.7), (6.4.8) and (6.4.9). The total number of equations is thus 30.

The unknowns are also 30: 12 generalized stresses (4 stresses $N_{\alpha\beta}$, 2 stresses $N_{\alpha 3}$, 4 stresses $M_{\alpha\beta}$ and 2 stresses $M_{\alpha 3}$), 12 generalized strains (4 strains $E_{\alpha\beta}$, 2 strains $E_{\alpha 3}$, 4 strains $K_{\alpha\beta}$ and 2 strains $K_{\alpha 3}$), and 6 generalized displacements (3 displacements U_i and 3 rotations Θ_i).

The boundaryconditions are 6 for each boundary point. Denoting by Γ_1 and Γ_2 the parts of the boundary where stresses and displacements are respectively prescribed, the boundary conditions are

$$\left.\begin{aligned} N_i &= \bar{N}_i \\ M_i &= \bar{M}_i \end{aligned}\right\} \text{ on } \Gamma_1 \tag{6.5.1}$$

$$\left.\begin{aligned} U_i &= \bar{U}_i \\ \Theta_i &= \bar{\Theta}_i \end{aligned}\right\} \text{ on } \Gamma_2 \tag{6.5.2}$$

6.6 Simplified Theories

A simplified theory is introduced if we make

$$E_{\alpha 3} = 0 \tag{6.6.1}$$

$$E_{12} = E_{21} \tag{6.6.2}$$

Such simplified two-dimensional theory can easily be connected with the simplified three-dimensional one if the displacement field in the shell, considered as a three-dimensional body, is assumed to be defined by

$$\vec{u}\left(\xi_1,\xi_2,\xi_3\right) = \vec{U}\left(\xi_1,\xi_2\right) - \vec{a}_3\xi \wedge \vec{\Theta}\left(\xi_1,\xi_2\right) \qquad (6.6.3)$$

$$\vec{\theta}\left(\xi_1,\xi_2,\xi_3\right) = \vec{\Theta}\left(\xi_1,\xi_2\right) \qquad (6.6.4)$$

Such would be the generalized displacements in the body if any straight segment normal to S would remain straight and inextensible after the deformation, and its rotation was $\vec{\Theta}$.

The strains corresponding to such displacement field are, by virtue of (5.3.8) and (5.3.9),

$$e_{\alpha i} = \frac{\vec{u}_{,\alpha}\cdot\vec{a}_i}{h_\alpha} - \vec{a}_\alpha \wedge \vec{a}_i\cdot\vec{\theta} = \frac{\vec{U}_{,\alpha}\cdot\vec{a}_i - \vec{a}_3\xi_3\wedge\vec{\Theta}_{,\alpha}\cdot\vec{a}_i}{h_\alpha} - \vec{a}_\alpha\wedge\vec{a}_i\cdot\vec{\Theta} \qquad (6.6.5)$$

$$e_{3i} = \frac{\vec{u}_{,3}\cdot\vec{a}_i}{h_3} - \vec{a}_3\wedge\vec{a}_i\cdot\vec{\theta} = -\vec{a}_3\wedge\vec{\Theta}\cdot\vec{a}_i - \vec{a}_3\wedge\vec{a}_i\cdot\vec{\Theta} = 0 \qquad (6.6.6)$$

$$k_{\alpha i} = \frac{\vec{\theta}_{,\alpha}\cdot\vec{a}_i}{h_\alpha} = \frac{\vec{\Theta}_{,\alpha}\cdot\vec{a}_i}{h_\alpha} \qquad (6.6.7)$$

$$k_{3i} = \frac{\vec{\Theta}_{,3}\cdot\vec{a}_i}{h_3} = 0 \qquad (6.6.8)$$

We conclude, considering (6.3.5) and (6.3.6), that

$$E_{\alpha i} = \left(e_{\alpha i}\right)_{\xi_3} = 0 \qquad (6.6.9)$$

$$K_{\alpha i} = \left(k_{\alpha i}\right)_{\xi_3} = 0 \qquad (6.6.10)$$

Assume now that we impose equality (5.6.1) on S. Such simplifying assumption is quite similar to the one used in the three-dimensional theory. The only difference is that equality (5.6.1) is postulated in the two-dimensional domain S and not, as before, in the three-dimensional domain V. There results

$$e_{ij} = e_{ji} \quad \text{on } S \tag{6.6.11}$$

and thus, by virtue of (6.6.6) and (6.6.9),

$$E_{\alpha 3} = (e_{\alpha 3})_{\xi_3 = 0} = (e_{3\alpha})_{\xi_3 = 0} = 0 \tag{6.6.12}$$

$$E_{12} = (e_{12})_{\xi_3 = 0} = (e_{21})_{\xi_3 = 0} = E_{21} \tag{6.6.13}$$

The simplifying assumptions of the two-dimensional theory correspond therefore to the simplifying assumptions of the three-dimensional theory.

We remark that, as $e_{\alpha 3}$ and $e_{3\alpha}$ both vanish, any straight segment normal to S not only remains straight and inextensible but also normal to S after the deformation. Such is Kirchkoff's assumption on which the theory of shells is usually built.

Assumptions (6.6.1) and (6.6.2) imply that the strain-displacement equations (6.3.5) can be written

$$E_{\alpha\beta} = \frac{1}{2}\left(\frac{\vec{U}_{,\alpha}\cdot\vec{a}_\beta}{k^0_\alpha} + \frac{\vec{U}_\beta\cdot\vec{a}_\alpha}{k^0_\beta}\right) = E_{\beta\alpha} \qquad (6.6.14)$$

and that the rotation components be given by

$$\Theta_1 = \frac{\vec{U}_{,2}\cdot\vec{a}_3}{h^0_2} \qquad (6.6.15)$$

$$\Theta_2 = -\frac{\vec{U}_{,1}\cdot\vec{a}_3}{h^0_1} \qquad (6.6.16)$$

$$\Theta_3 = \frac{1}{2}\left(\frac{\vec{U}_{,1}\cdot\vec{a}_2}{h_1} - \frac{\vec{U}_{,2}\cdot\vec{a}_2}{h_2}\right) \qquad (6.6.17)$$

Simplifying assumption (6.6.1) would be exact if constant ν were indefinitely large. On the other hand, assumption (6.6.2) would be exact, if the same would happen to $\mathcal{P}$.

We observe that assumptions (6.6.1) and (6.6.2) are independent from each other. We may have thus three simplified theories according to our adopting (6.6.1), (6.6.2) or both.

6.7 Balance of Equations in the Simplified Theories

Let us consider first the theory based on the simplifying assumption (6.6.1), i.e., on neglecting the transverse shear deformability.

Adopting (6.6.1) obliges to disregarding the stress-strain equation (6.4.7). The number of the stress-strain equations decreases thus to 10. The number of independent non-trivial strain-displacement equations is also 10, involving 10 generalized strains $\left(E_{\alpha\beta}, K_{\alpha\beta}, K_{\alpha 3}\right)$ and 4 generalized displacements $\left(U_i , \Theta_3\right)$.

The number of generalized stress involved in the stress-strain equations is also 10 $\left(N_{\alpha\beta} , M_{\alpha\beta} , M_{\alpha i}\right)$.

The missing 4 equilibrium equations can be established by eliminating N_{13} and N_{23} between (6.2.14) and (6.2.15). The total number of equations and unknowns is thus 24.

Let us consider the theory based on assumption (6.6.2). Stress-strain equations (6.4.6) must be replaced by

$$\frac{N_{\alpha\beta}+N_{\beta\alpha}}{2} = 2\bar{\mu}\,\frac{E_{\alpha\beta}+E_{\beta\alpha}}{2} + \bar{\lambda}\,E\,\delta_{\alpha\beta} \tag{6.7.1}$$

The number of independent stress-strain equations is thus 11.

We have also 11 strain-displacement equations.

The independent unknowns are: 11 generalized

strains (because $E_{\alpha\beta} = E_{\beta\alpha}$), 11 generalized stresses $\left(\frac{N_{\alpha\beta}+N_{\beta\alpha}}{2},\right.$ $\left. N_{\alpha 3}, M_{\alpha i}\right)$ and 5 generalized displacements $\left(U_i, \Theta_\alpha\right)$.

The 5 missing equilibrium equations can be established by eliminating the antisymmetric part of $N_{\alpha\beta}$ between (6.2.14) and (6.2.15). The total number of equations and unknowns in the second simplified theory is thus 27.

If (6.6.1) and (6.6.2) are both adopted, the number of equations and unknowns decreases to 9 + 9 + 3 = 21.

6.8 Boundary Conditions in the Simplified Theories

The simplified assumption (6.6.2) does not affect the boundary conditions. It happens otherwise with assumption (6.6.1). In case assumption (6.6.1) is adopted, the displacement boundary conditions become the following five

$$U = \vec{\bar{U}} \tag{6.8.1}$$

$$\vec{\Theta} \wedge \vec{n}^{\,0} = \vec{\bar{R}} \tag{6.8.2}$$

$\vec{\bar{U}}$ and $\vec{\bar{R}}$ being vectors prescribed on Γ_2 .

The stress-boundary conditions can be determined by considering the work done by the forces applied to the boundary when the displacement and rotation fields are given virtual variations $\delta\vec{U}$ and $\delta\vec{\Theta}$.

Considering that $\delta U = 0$ on Γ_2 , we have

$$\delta\tau = \int_{\Gamma}\left(\vec{N}\cdot\delta\vec{U} + \vec{M}\cdot\delta\vec{\Theta}\right)d\Gamma = \int_{\Gamma_1}\left(\vec{N}\cdot\delta\vec{U} + \vec{M}\cdot\delta\vec{\Theta}\right)d\Gamma$$

(6.8.3)

But

(6.8.4) $$\vec{M}\cdot\delta\vec{\Theta} = M_n\delta\Theta_n + \vec{M}\wedge\vec{n}\cdot\delta\vec{\Theta}\wedge\vec{n}$$

On the other hand, (6.6.15) and (6.6.16) imply

(6.8.5) $$\delta\Theta_n = \frac{\delta\vec{U}_{,\Gamma}\cdot\vec{a}_3}{h^0_\Gamma}$$

Thus

$$\int_{\Gamma_1} M_n\frac{\delta\vec{U}_{,\Gamma}\cdot\vec{a}_3}{h^0_\Gamma}d\Gamma = \int_{\Gamma_1}\left(\frac{M_n\delta\vec{U}\cdot\vec{a}_3}{h^0_\Gamma}\right)_{,\Gamma}d\Gamma - \int_{\Gamma_1}\delta\vec{U}\cdot\left(\frac{M_n\vec{a}_3}{h^0_\Gamma}\right)_{,\Gamma}d\Gamma$$

(6.8.6)

But

(6.8.7) $$\int_{\Gamma_1}\left(M_n\frac{\delta\vec{U}\cdot\vec{a}_3}{h^0_\Gamma}\right)_{,\Gamma}d\Gamma = \left[\frac{M_n\delta\vec{U}\cdot\vec{a}_3}{h^0_\Gamma}\right]_{\Gamma_1^1}^{\Gamma_1^2}$$

in which Γ_1^1 and Γ_1^2 denote the two ends of Γ_1.

As such ends are also points of Γ_2 and $\delta\vec{U}$ vanishes on Γ_2, there follows that the right-hand side of (6.8.7) vanishes. It would vanish also if only stresses were prescribed on the boundary, i.e. if Γ_1 would coincide with Γ, because then, the two ends of Γ_1 would coincide.

Introducing (6.8.6), (6.8.3) becomes

$$\delta\tau = \int_{\Gamma_1}\left[\vec{N}\cdot\delta\vec{U} + \vec{M}\wedge\vec{n}^{\circ}\cdot\delta\vec{\Theta}\wedge\vec{n}^{\circ} - \left(\frac{M_n\vec{a}_3}{h^0_\Gamma}\right)_{,\Gamma}\delta\vec{U}\right]d\Gamma =$$

$$= \int_{\Gamma_1}\left\{\left[\vec{N} - \left(\frac{M_n\vec{a}_3}{h^0_\Gamma}\right)_{,\Gamma}\right]\delta\vec{U} + \vec{M}\wedge\vec{n}^{\circ}\cdot\delta\vec{\Theta}\wedge\vec{n}^{\circ}\right\}d\Gamma . \quad (6.8.8)$$

The stress boundary conditions are thus

$$\vec{N} - \left(\frac{M_n\vec{a}_3}{h^0_\Gamma}\right)_{,\Gamma} = \vec{\bar{S}} \quad (6.8.9)$$

$$\vec{M}\wedge\vec{n}^{\circ} = \vec{\bar{T}} \quad (6.8.10)$$

$\vec{\bar{S}}$ and $\vec{\bar{T}}$ being vectors prescribed on Γ_1 .

6.9 The Classical Theory of Shells

The classical theory of shells is, like classical elasticity, based on the assumption that the couple-stresses μ_{ij} vanish in the shell. We know that this happens if constants ψ , χ and φ are equal to zero.

Introducing such assumption in (6.2.7) and (6.2.8) there results

$$\vec{M} = \int_{-t/2}^{t/2} \lambda\,\vec{a}_3\,\xi_3\wedge\vec{\sigma}\,d\xi_3 \quad (6.9.1)$$

$$\vec{M}_\alpha = \int_{-t/2}^{t/2} \lambda_{\bar{\alpha}}\,\vec{a}_3\,\xi_3\wedge\vec{\sigma}_\alpha\,d\xi_3 \quad (6.9.2)$$

The important point is that M_3 and $M_{\alpha 3}$ are now equal to zero. The number of generalized stresses decreases thus to 10 $(N_{\alpha i}, M_{\alpha\beta})$.

As $M_{\alpha 3}$ is equal to zero in any case, the elastic constant τ must also be zero. This means that $K_{\alpha 3}$ is arbitrary and that the corresponding strain-displacement equations loose their meaning.

We also remark that the strain-displacement equations were deduced from (6.3.2) under the assumption that the components of $\vec{N}_\alpha$ and $\vec{M}_\alpha$ were arbitrary. If $M_{\alpha 3}$ is always equal to zero, (6.3.2) cannot yield more than (6.3.5) and

$$K_{\alpha\beta} = \frac{\Theta_{,\alpha}}{h^0_\alpha} \cdot \vec{a}_\beta \tag{6.9.3}$$

There are thus no more than 10 generalized strains $(E_{\alpha i}, K_{\alpha\beta})$ and 10 strain-displacement equations in the classical theory of shells. 10 is also the number of stress-strain relations.

The total number of equations and unknowns is thus $10 + 10 + 6 = 26$. The number of boundary conditions is 5, because M_3 always vanishes. The boundary conditions are thus

$$\left.\begin{aligned} N_i &= \bar{N}_i \\ M_\alpha &= \bar{M}_\alpha \end{aligned}\right\} \text{ on } \Gamma_1 \tag{6.9.4}$$

$$\left.\begin{array}{l} U_i = \bar{U}_i \\ \Theta_\alpha = \bar{\Theta}_\alpha \end{array}\right\} \text{ on } \Gamma_2 \tag{6.9.5}$$

The general classical theory of shells may be simplified in several ways.

We can neglect for instance the transverse shear deformability, i.e. assume (6.6.1). The total number of unknowns and equations decreases then by 6, as it was shown in 6.7. On the other hand, assuming (6.6.2) makes such number decrease by 3.

Another important simplification consists in assuming the ratios $\frac{t}{R_\alpha}$ to be so small that magnitudes λ and λ_α can be made constant and equal to unity. Then, if on the other hand, $\vec{g}$ is assumed to vanish, and thus, $\sigma_{\alpha\beta}$ to be equal to $\sigma_{\beta\alpha}$, we have, by virtue of (6.2.4) and (6.9.2)

$$N_{\alpha\beta} = N_{\beta\alpha} \tag{6.9.6}$$

$$M_{11} = -M_{22} \tag{6.9.7}$$

Equalities (6.9.6) and (6.9.7) generally violate the equilibrium equation (6.2.15).

Indeed, one of the scalar equations (6.2.15) is equivalent to

$$\frac{M_{11}}{R_1^0} + \frac{M_{22}}{R_2^0} + N_{12} - N_{21} + G_3 = 0 \tag{6.9.8}$$

But, as $g = 0$

$$\vec{G} = \int_{-t/2}^{t/2} \lambda_1 \lambda_2 \vec{a}_3 \wedge \xi_3 \wedge \vec{f} \, d\xi_3 \tag{6.9.9}$$

and thus

$$G_3 = 0 \tag{6.9.10}$$

Then, (6.9.6), (6.9.7), (6.9.8) can hold simultaneously only if

$$\frac{1}{R_1^0} - \frac{1}{R_2^0} = 0 \tag{6.9.11}$$

which generally is not true.

As the components of $\vec{N}_\alpha$ and $\vec{M}_\alpha$ are no more arbitrary, (6.3.2) cannot yield more than

$$\frac{E_{\alpha\beta} + E_{\beta\alpha}}{2} = \frac{1}{2}\left(\frac{\vec{U}_{,\alpha} \cdot \vec{a}_\beta}{h_\alpha^0} + \frac{\vec{U}_{,\beta} \cdot \vec{a}_\alpha}{h_\beta^0}\right) \tag{6.9.12}$$

$$E_{\alpha 3} = \vec{U}_{,\alpha} \cdot \vec{a}_3 - \vec{a}_\alpha \wedge \vec{a}_3 \cdot \vec{\Theta} \tag{6.9.13}$$

$$\frac{K_{11} - K_{22}}{2} = \frac{1}{2}\left(\frac{\vec{\Theta}_{,1}}{h_1^0} \cdot \vec{a}_1 - \frac{\vec{\Theta}_{,2}}{h_2^0} \cdot \vec{a}_2\right) \tag{6.9.14}$$

$$K_{12} = \frac{\vec{\Theta}_{,1} \cdot \vec{a}_2}{h_1^0} \tag{6.9.15}$$

$$K_{21} = \frac{\vec{\Theta}_{,2} \cdot \vec{a}_1}{h_2^0} \tag{6.9.16}$$

It is possible to show that the rotation component Θ_3 is not involved in equations (6.9.12), (6.9.13), (6.9.15) and (6.9.16). This is clear for equations (6.9.12) and (6.9.13) and can also be made clear for equations (6.9.15) and (6.9.16) if it is remarked that, by virtue of (6.1.2)

$$\left(\Theta_3 \vec{a}_3\right)_{,\alpha} \cdot \vec{a}_\beta = \Theta_{3,\alpha} \vec{a}_3 \cdot \vec{a}_\beta + \Theta_3 \vec{a}_{3,\alpha} \cdot \vec{a}_\beta = \Theta_3 \frac{h^0_\alpha}{R^0_\alpha} \vec{a}_\alpha \cdot \vec{a}_\beta = \Theta_3 \frac{h^0_\alpha}{R^0_\alpha} \delta_{\alpha\beta} \tag{6.9.17}$$

Let us examine now what happens with equation (6.9.14).

Introducing (6.9.17) into (6.9.14), there results

$$\frac{K_{11} - K_{22}}{2} = \frac{1}{2} \sum_\alpha \left[\frac{(\Theta_\alpha \vec{a}_\alpha)_{,1}}{h^0_1} \cdot \vec{a}_1 - \frac{(\Theta_\alpha \vec{a}_\alpha)_{,2}}{h^0_2} \cdot \vec{a}_2 \right] + \frac{1}{2} \Theta_3 \left(\frac{1}{R^0_1} - \frac{1}{R^0_2} \right) \tag{6.9.18}$$

If we neglect the term containing Θ_3, the strain-displacement equation (6.9.14) will become only approximate. It will be exact only if (6.9.11) is true. This is similar to what happens with the equilibrium equation (6.9.8).

As one of the equilibrium equations must be disregarded and the terms containing Θ_3 are neglected in the strain-displacement equations, both the number of equilibrium equations and generalized displacements become 5.

The stress-strain equations become

$$N_{\alpha\beta} = 2 \bar{\mu} \frac{E_{\alpha\beta} + E_{\beta\alpha}}{2} + \bar{\lambda} E \delta_{\alpha\beta} \tag{6.9.19}$$

(6.9.20) $$N_{\alpha 3} = 2\bar{\nu} E_{\alpha 3}$$

(6.9.21) $$M_{11} = -M_{22} = 2\bar{\psi}\,\frac{K_{11}-K_{22}}{2}$$

(6.9.22) $$M_{12} = 2\bar{\psi}\,\frac{K_{12}+K_{21}}{2} + 2\bar{\chi}\,\frac{K_{12}-K_{21}}{2}$$

(6.9.23) $$M_{21} = 2\bar{\psi}\,\frac{K_{12}+K_{21}}{2} + 2\bar{\chi}\,\frac{K_{12}-K_{21}}{2}$$

Equation (6.9.21) was obtained by averaging the equations for M_{11} and M_{22} and equation (6.9.19) by averaging the equations for $N_{\alpha\beta}$ and $N_{\beta\alpha}$.

In case the stresses $\sigma_{\alpha\beta}$ are assumed to distribute linearly in the thickness of the shell, and the stresses $\sigma_{\alpha 3}$ are assumed to distribute parabolically vanishing for $\xi_3 = \pm \frac{t}{2}$, the elastic constants $\bar{\mu}, \bar{\lambda}, \bar{\nu}, \bar{\psi}$ and $\bar{\chi}$ can be shown to have the following well-known expressions in terms of λ and μ:

(6.9.24) $$\bar{\mu} = \mu t$$

(6.9.25) $$\bar{\lambda} = \lambda t$$

(6.9.26) $$\bar{\nu} = \frac{5}{12}\mu t$$

(6.9.27) $$\bar{\psi} = \frac{\mu t^3}{12}$$

(6.9.28) $$\bar{\chi} = \frac{3\lambda + 2\mu}{\lambda + 2\mu} \cdot \frac{\mu t_3}{12} .$$

6.10 Theory of Membranes

Assuming

$$\sum_{\alpha} \left(h^{0}_{\bar{\alpha}} \vec{M}_{\alpha}\right)_{,\alpha} + h^{0} \vec{G} = \vec{0} \tag{6.10.1}$$

equation (6.2.15) becomes

$$\sum_{\alpha} \vec{a}^{0}_{\alpha} \wedge \vec{N}_{\alpha} = \vec{0} \tag{6.10.2}$$

which is equivalent to

$$N_{23} \vec{a}_1 - N_{13} \vec{a}_2 + \left(N_{12} - N_{21}\right) \vec{a}_3 = \vec{0} \tag{6.10.3}$$

Equation (6.10.3) implies $N_{\alpha 3} = 0$ and $N_{12} = N_{21}$.

Introducing such conditions in (6.2.13), we obtain the 3 following equations involving the 3 unknowns $N_{11}, N_{12} = N_{21}$ and N_{22}:

$$\left(h^{0}_{2} N_{11}\right)_{,1} + \left(h^{0}_{1} N_{21}\right)_{,2} + h^{0}_{1,2} N_{12} - h^{0}_{2,1} N_{22} + h^{0}_{1} h^{0}_{2} F_1 = 0 \tag{6.10.4}$$

$$\left(h^{0}_{2} N_{12}\right)_{,1} + \left(h^{0}_{1} N_{22}\right)_{,1} + h^{0}_{2,1} N_{21} - h^{0}_{1,1} N_{11} + h^{0}_{1} h^{0}_{2} F_2 = 0 \tag{6.10.5}$$

$$-\frac{N_{11}}{R_1} - \frac{N_{22}}{R_2} + F_3 = 0 \tag{6.10.6}$$

Any solution of such equations, together with $M_{\alpha i} = 0$, can be interpreted as a particular solution of the shell equilibrium equations.

It will be the exact solution if $\vec{G} = \vec{0}$ and the elastic constants

$\bar{\psi}$, $\bar{\chi}$, $\bar{\varphi}$ and $\bar{\tau}$ are equal to zero because then, the moments $M_{\alpha i}$ vanish and equation (6.10.1) is satisfied.

We remark that such constants cannot be exactly equal to zero without $\bar{\mu}$ and $\bar{\lambda}$ being equal to zero, within the frame of the classical theory, in which the relations (6.9.24-28) are assumed to hold.

On the other hand, as

$$N_{12} = \int_{-t/2}^{t/2} \lambda_2 \sigma_{12} \, d\xi_3 \tag{6.10.7}$$

$$N_{21} = \int_{-t/2}^{t/2} \lambda_1 \sigma_{21} \, d\xi_3 \tag{6.10.8}$$

N_{12} cannot be equal to N_{21} in the classical theory, unless $\lambda_1 = \lambda_2$, i.e. unless $R_1 = R_2$. Such difficulty disappears in the general theory, however, in which σ_{12} can be different from σ_{21} and thus

$$\lambda_2 \sigma_{12} = \lambda_1 \sigma_{21} \tag{6.10.9}$$

even if λ_1 is different from λ_2.

6.11 The Plane Theory

If the domain is plane $(R_1^o = R_2^o = \infty)$ and the system of coordinates is cartesian, the equilibrium equations (6.2.14) and (6.2.15) become

$$N_{11,1} + N_{21,2} + F_1 = 0 \tag{6.11.1}$$

$$N_{12,1} + N_{22,2} + F_2 = 0 \tag{6.11.2}$$

$$N_{13,1} + N_{23,2} + F_3 = 0 \tag{6.11.3}$$

$$M_{11,1} + M_{21,2} + N_{23} + G_1 = 0 \tag{6.11.4}$$

$$M_{12,1} + M_{22,2} - N_{13} + G_2 = 0 \tag{6.11.5}$$

$$M_{13,1} + M_{23,2} + N_{12} - N_{21} + G_3 = 0 . \tag{6.11.6}$$

The strain-displacement equations take the form

$$E_{11} = U_{1,1} \tag{6.11.7}$$

$$E_{12} = U_{2,1} - \Theta_3 \tag{6.11.8}$$

$$E_{13} = U_{3,1} + \Theta_2 \tag{6.11.9}$$

$$E_{21} = U_{1,2} + \Theta_3 \tag{6.11.10}$$

$$E_{22} = U_{2,2} \tag{6.11.11}$$

$$E_{23} = U_{3,2} - \Theta_1 \tag{6.11.12}$$

$$K_{\alpha\beta} = \Theta_{\beta,\alpha} \tag{6.11.13}$$

$$K_{\alpha 3} = \Theta_{3,\alpha} \tag{6.11.14}$$

These equations and the stress-strain relations (6.4.6-9) can be split into two groups forming two separate theories.

The first one is the Cosserat plane theory (plane stresses), characterized by equilibrium equations (6.11.1), (6.11.2) and (6.11.6), by strain-displacement equations (6.11.7), (6.11.8), (6.11.10), (6.11.11) and (6.11.14) and by stress-strain equations (6.4.6) and (6.4.9).

The second one is the theory of plates which consists of equilibrium equations (6.11.3), (6.11.4) and (6.11.5) strain-displacement equations (6.11.9), (6.11.12) and (6.11.13) and stress-strain equations (6.4.7) and (6.4.8).

The boundary conditions for the first theory are

$$\left.\begin{aligned} N_\alpha &= \bar{N}_\alpha \\ M_3 &= \bar{M}_3 \end{aligned}\right\} \text{ on } \Gamma_1 \tag{6.11.15}$$

$$\left.\begin{aligned} U_\alpha &= \bar{U}_\alpha \\ \Theta_3 &= \bar{\Theta}_3 \end{aligned}\right\} \text{ on } \Gamma_2 \tag{6.11.16}$$

and for the second theory

$$\left.\begin{aligned} N_3 &= \bar{N}_3 \\ M_\alpha &= \bar{M}_\alpha \end{aligned}\right\} \text{ on } \Gamma_1 \tag{6.11.17}$$

$$\left.\begin{aligned} U_3 &= \bar{U}_3 \\ \Theta_\alpha &= \bar{\Theta}_\alpha \end{aligned}\right\} \text{ on } \Gamma_2 \tag{6.11.18}$$

Cosserat plane theory can be simplified by using assumption (6.6.2). Acting as in the general case, we obtain a simplified theory consisting of the following equations.

- equilibrium equations

$$\sum_\alpha \left(\frac{N_{1\alpha}+N_{\alpha 1}}{2}\right)_{,\alpha} + \frac{1}{2}\sum_\alpha M_{\alpha 3,\alpha 2} + F_1 + \frac{1}{2} G_{3,2} = 0 \tag{6.11.19}$$

$$\sum_\alpha \left(\frac{N_{2\alpha}+N_{\alpha 2}}{2}\right)_{,\alpha} - \frac{1}{2}\sum_\alpha M_{\alpha 3,\alpha 3} + F_2 + \frac{1}{2} G_{3,1} = 0 \tag{6.11.20}$$

- strain- displacement equations

$$\frac{E_{\alpha\beta}+E_{\beta\alpha}}{2} = \frac{1}{2}\left(U_{\alpha,\beta}+U_{\beta,\alpha}\right) \tag{6.11.21}$$

$$K_{\alpha\beta} = \frac{1}{2}\left(U_{2,1\alpha}-U_{1,2\alpha}\right) \tag{6.11.22}$$

- stress-strain equations

(6.11.23) $$\frac{N_{\alpha\beta}+N_{\beta\alpha}}{2} = 2\,\bar{\mu}\,\frac{E_{\alpha\beta}+E_{\beta\alpha}}{2} + \bar{\lambda}\,E\,\delta_{\alpha\beta}$$

(6.11.24) $$M_{\alpha 3} = 2\,\bar{\tau}\,K_{\alpha 3}$$

Such equations are supplemented by boundary equations (6.11.15) and (6.11.16).

Making $\bar{\tau}=0$ and assuming $G_3=0, N_{12}$ becomes equal to N_{21} by virtue of (6.11.6) and the classical theory (plane stress) is obtained.

The theory of plates can be simplified by using assumption (6.6.1) which gives

(6.11.25) $$\Theta_2 = -U_{3,1}$$

(6.11.26) $$\Theta_1 = U_{3,2}\,.$$

We obtain the following equations

- equilibrium equations

(6.11.27) $$M_{12,11} + \left(M_{22}-M_{11}\right)_{,12} - M_{21,22} + G_{2,1} - G_{1,2} + F_3 = 0$$

- strain-displacement equations

(6.11.28) $$K_{\alpha 1} = U_{3,2\alpha}$$

(6.11.29) $$K_{\alpha 2} = -U_{3,1\alpha}$$

which lead to $K_{11} = -K_{22}$ and thus to $K = 0$.

- stress-strain equations

$$\frac{M_{11} - M_{22}}{2} = 2\,\bar{\psi}\,K_{11} = -2\,\bar{\chi}\,K_{22} \qquad (6.11.30)$$

$$M_{12} = 2\,\bar{\psi}\,\frac{K_{12} + K_{21}}{2} + 2\,\bar{\chi}\,\frac{K_{12} - K_{21}}{2} \qquad (6.11.31)$$

$$M_{21} = 2\,\bar{\psi}\,\frac{K_{12} + K_{21}}{2} + 2\,\bar{\chi}\,\frac{K_{12} - K_{21}}{2} \qquad (6.11.32)$$

- boundary conditions

$$\left.\begin{aligned} N_3 - M_{n,\Gamma} &= \bar{S}_3 \\ M_\alpha &= \bar{M}_\alpha \end{aligned}\right\} \text{ on } \Gamma_1 \qquad (6.11.33)$$

$$\left.\begin{aligned} U_3 &= \bar{U}_3 \\ \Theta_\alpha &= \bar{\Theta}_\alpha \end{aligned}\right\} \text{ on } \Gamma_2 . \qquad (6.11.34)$$

Assuming $M_{11} = -M_{22}$, we obtain the theory of plates under its classical form. The stress-strain equations (6.11.30) become

$$M_{11} = -M_{22} = 2\,\bar{\psi}\,K_{11} = -2\,\bar{\psi}\,K_{22} . \qquad (6.11.35)$$

Introducing (6.11.31), (6.11.32) and (6.11.35) in (6.11.27), and using (6.11.28) and (6.11.29), we obtain Lagrange's equation under the form

(6.11.36) $$\nabla^4 U_3 = \frac{1}{\bar{\psi}+\bar{\chi}}\left(F_3 + G_{2,1} - G_{1,2}\right)$$

If relations (6.9.27) and (6.9.28) are assumed to hold,

(6.11.37) $$\bar{\psi}+\bar{\chi} = \frac{\mu t_3}{12}\left(\frac{3\lambda+2\mu}{\lambda+\mu}+1\right) = \frac{E}{1-\nu^2}\,\frac{t^3}{12}$$

Equation (6.11.36) takes then the more usual form

(6.11.38) $$\nabla^4 U_3 = \frac{1-\nu^2}{\frac{Et^3}{12}}\left(F_3 + G_{2,1} - G_{1,2}\right).$$

Let us remark, to close our considerations about plates, that two tensors can be formed with magnitudes $M_{\alpha\beta}$ in case the couple-stresses $\mu_{\alpha\beta}$ are assumed to vanish. Vector $\vec{M}_\alpha$ has then indeed the following expression

(6.11.39) $$\vec{M}_\alpha = \int_{-t/2}^{t/2} \xi_3\, \vec{a}_3 \wedge \vec{\sigma}_3\, d\xi_3$$

The two tensors are $\left[M_{\alpha\beta}\right]$ and $\left[\int_{-t/2}^{t/2} \xi_3 \sigma_{\alpha\beta}\, d\xi_3\right]$

The matrix of the first is

(6.11.40) $$\underline{M} = \begin{bmatrix} M_{11} & M_{12} \\ M_{21} & M_{22} \end{bmatrix}$$

and the matrix of the second is

$$\underline{M}' = \begin{bmatrix} M_{12} & -M_{11} \\ M_{22} & -M_{21} \end{bmatrix} \qquad (6.11.41)$$

The second tensor is usually the only one considered in the literature [37] not only because of its more obvious connexion with the ordinary stress tensor but also because of its being symmetric in case the $\mu_{\alpha i}$ are equal to zero.

6.12 Closing Remarks

It results clearly, from the preceding sections that a two-dimensional model can be established independently of the three-dimensional one (°).

A physical assumption was subtly introduced when the two-dimensional strain energy deformation density was made an exclusive function of the $E_{\alpha i}$ and $K_{\alpha i}$.

An analogous assumption is however implicit

(°) And the same could have been done for the one-dimensional model.

in classical elasticity when it is stated that the three-dimensional strain energy density is an exclusive function of the ordinary strains e_{ij} . The approximate nature of such assumption is quite clear if it is observed that classical elasticity is a particular case of Cosserat's theory, which in turn is a particular case of the multipolar theories (and so on...).

The three-dimensional model is privileged [38] not because it is more consistent in itself but because it is believed to provide a better simulation of the behaviour of bodies. For this reason, and not for any other, it is usually looked at as the model which generates all the others.

7. THE DISCRETE MODEL

7.1 The Equations

The discrete model of the theory of structures fits into the general scheme presented in Chapter 4.

Usually, the discrete model is not presented independently but connected to continuous models from which it is generated. However, there is no reason for not presenting the discrete model independently, as the two and three-dimensional models considered in the preceeding chapter.

While the two and three-dimensional continuous models were respectively referred to two and three-dimensional domains, the discrete model is referred to what we may call a discrete domain. The structure is indeed decomposed into parts called elements and the set of such parts is what we call the discrete domain to which the discrete model is referred.

Integrations on the domain are thus replaced by summations over the set of elements.

The continuous models involve fields of stresses, strains and displacements. Values of the stresses, strains and displacements are associated to each point of the domain. In the discrete model, as the domain consists of a discrete set of elements, each field is replaced by a set of vectors associated to the set of elements. The stress, strain and displacement vectors associated to a general element will be respectively denoted by $\underline{\sigma}^e, \underline{e}^e, \underline{u}^e$.

A complete analogy could be established with continuous models if the displacement vector were partitioned into an internal displacement vector , $\underline{u}_i^e$, and an external displacement vector.

(7.1.1) $$\underline{u}^e = \begin{Bmatrix} \underline{u}_i^e \\ \underline{u}_i^e \end{Bmatrix}$$

and internal and external forces $\underline{f}^e$ and $\underline{p}^e$ corresponding to such displacements were assumed to act on the element.

Two kinds of equilibrium equations could be considered then,

(7.1.2) $$\underline{E}^e \underline{\sigma}^e = \underline{f}^e$$

(7.1.3) $$\underline{N}^e \underline{\sigma}^e = \underline{p}^e$$

corresponding to equations (4.2.1) and (4.2.4) respectively.

No loss of generality is introduced however if such distinction is not made and all the displacements and forces are assumed to be external. The only equilibrium equations to be considered are then represented by (7.1.3).

The following equations can be written then for each element

(7.1.4) equilibrium equations $\underline{N}^e \underline{\sigma}^e = \underline{p}^e$

strain-displacement equations $\underline{e}^e = \underline{e}^{oe} + \underline{D}^e \underline{u}^e$ (7.1.5)

stress-strain equations $$\underline{s}^e = \frac{\partial U^e}{\partial \underline{e}^e} \qquad (7.1.6)$$

$\underline{D}^e$ is now a simple matrix and not a differential operator. $\underline{U}^e$ is the strain energy of the element, a definite positive function of the strains. $\underline{p}^e$ is the vector of the forces acting on element e.

Matrices $\underline{D}^e$ and $\underline{N}^e$ must be such that the work principle holds. The work principle has the form

$$\underline{s}^{eT}(\underline{D}^e \underline{u}^e) = (\underline{N}^e \underline{s}^e)^T \underline{u}^e \qquad (7.1.7)$$

or

$$\underline{s}^{eT} \underline{e}^e = \underline{s}^{eT} \underline{e}^{oT} + \underline{p}^{eT} \underline{u}^e \qquad (7.1.7')$$

where $\underline{p}^e$ is associated to $\underline{s}^e$ through (7.1.4), and $\underline{e}^e$ to $\underline{e}^{oe}$ and $\underline{u}^e$ through (7.1.5). Equations (7.1.7) and (7.1.7') correspond to (4.2.13) and (4.2.13').

As equations (7.1.7) must be valid for any vectors $\underline{s}^e$ and $\underline{u}^e$ matrices $\underline{D}^e$ and $\underline{N}^e$ must be related by

$$\underline{D}^{eT} = \underline{N}^e \qquad (7.1.8)$$

Equation (7.1.8) corresponds to (4.7.8).

Compatibility between elements is defined by prescribing displacement discontinuities in terms of the vectors

$\underline{u}^e$ corresponding to the different elements. Such compatibility conditions may easily be introduced if the displacements of each element are compared with a vector of displacement of the system $\underline{U}$. Then we write

$$\underline{u}^e - \underline{A}^e \underline{U} = \bar{\underline{g}}^e \tag{7.1.9}$$

where vector $\bar{\underline{g}}^e$ contains the displacement discontinuities, i.e. the difference between the displacements of the element and the corresponding displacements of the system. Equation (7.1.9) corresponds to (4.2.11).

As $\bar{\underline{g}}^e$ will be assumed to vanish in the sequel, equation (7.1.9) will be particularized to

$$\underline{u}^e = \underline{A}^e \underline{U}\,. \tag{7.1.10}$$

External forces corresponding to displacements $\underline{U}$ are supposed to act on the element boundaries. The equations which express the equilibrium between such forces and the forces which act on the elements can easily be established by considering the invariance of work which implies

$$\sum_e u^{e^T} \underline{p}^e = \underline{U}^T \underline{F} \tag{7.1.11}$$

Introducing (7.1.10) into (7.1.11), there results

$$\underline{U}^T \sum_e \underline{A}^{e^T} \underline{p}^e = \underline{U}^T \underline{F} \tag{7.1.12}$$

and then

$$\sum_e \underline{A}^{e^T} \underline{p}^e = \underline{F} . \qquad (7.1.13)$$

Equation (7.1.13) corresponds to (4.2.12).

The strain energy associated with the system is

$$U = \sum_e U^e \qquad (7.1.14)$$

and the complementary energy is

$$U^* = \sum_e U^{e*} \qquad (7.1.15)$$

where

$$U^{e*} = \underline{s}^{e^T} \underline{e}^e - U^e \qquad (7.1.16)$$

Differentiation of (7.1.16) with respect to $\underline{e}^e$ yields

$$\frac{\partial U^{e*}}{\partial \underline{e}^e} = \underline{0} \qquad (7.1.17)$$

Differentiating with respect to the stresses there results

$$\frac{\partial U^{*e}}{\partial \underline{s}^e} = \underline{e}^e \qquad (7.1.18)$$

Linearity will be achieved if U^e is a quadratic function of the strains

$$U^e = \frac{1}{2} \underline{e}^{e^T} \underline{H}^e \underline{e}^e \qquad (7.1.19)$$

where $\underline{H}^e$ is symmetric and positive definite.

The stress-strain equations become then

(7.1.20) $$\underline{s}^e = \underline{H}^e \underline{e}^e .$$

Linearity will be assumed in what follows.

The solution of the system of equations (7.1.4), (7.1.5), (7.1.10), (7.1.13) and (7.1.20) can greatly be simplified if the number of unknowns is reduced by elimination. Such elimination permits to formulate the problem exclusively in terms of displacements (displacement method) or forces (force method.

Let us consider the displacement method first.

Introducing (7.1.20) into (7.1.3) we obtain the stress-displacement equations

(7.1.21) $$\underline{s}^e = \underline{H}^e \underline{e}^{oe} + \underline{H}^e \underline{D}^e \underline{u}^e .$$

Introducing (7.1.21) into (7.1.4), there results

(7.1.22) $$\underline{N}^e \underline{H}^e \underline{e}^{oe} + \underline{N}^e \underline{H}^e \underline{D}^e \underline{u}^e = \underline{p}^e .$$

Introduce now (7.1.22) into (7.1.13), we obtain

(7.1.23) $$\sum_e \underline{A}^{eT} \underline{N}^e \underline{H}^e \underline{e}^{oe} + \sum_e \underline{A}^{eT} \underline{N}^e \underline{H}^e \underline{D}^e \underline{u}^e = \underline{F} .$$

As the stresses associated with the initial strains are supposed to equilibrate vanishing external forces,

$$\sum_e \underline{A}^{eT} \underline{N}^e \underline{H}^e \underline{e}^{oe} = \underline{0} \qquad (7.1.24)$$

(7.1.24) and the introduction of (7.1.10) into (7.1.23) yield finally

$$\underline{K}\,\underline{U} = \underline{F} \qquad (7.1.25)$$

where

$$\underline{K} = \sum_e \underline{A}^{eT} \underline{N}^e \underline{H}^e \underline{D}^e \underline{A}^e \qquad (7.1.26)$$

Equation (7.1.25) corresponds to equation (4.2.25).

$\underline{K}$ is called the stiffness matrix of the system. $\underline{K}$ is a symmetric matrix because $\underline{H}^e$ is also symmetric and, by virtue of (7.1.8), (7.1.26) can be transformed into

$$\underline{K} = \sum_e \underline{A}^{eT} \underline{D}^{eT} \underline{H}^e \underline{D}^e \underline{A}^e . \qquad (7.1.27)$$

$\underline{K}$ is however a singular matrix because $\underline{U}$ includes rigid body displacements which correspond to $\underline{F} = \underline{0}$.

Let vectors $\underline{U}$ and $\underline{F}$ be partitioned according to the magnitudes which are given and those which are not

$$\underline{U} = \begin{Bmatrix} \underline{U}_1 \\ \underline{\bar{U}}_2 \end{Bmatrix} \qquad (7.1.28)$$

$$\underline{F} = \begin{Bmatrix} \underline{\bar{F}}_1 \\ \underline{F}_2 \end{Bmatrix} \qquad (7.1.29)$$

Partitioning $\underline{K}$ in conformity with (7.1.28) and (7.1.29), there results

$$\underline{K}_{11}\underline{U}_1 + \underline{K}_{12}\underline{\bar{U}}_2 = \underline{\bar{F}}_1 \tag{7.1.30}$$

$$\underline{K}_{21}\underline{U}_1 + \underline{K}_{22}\underline{\bar{U}}_2 = \underline{\bar{F}}_2 . \tag{7.1.31}$$

From (7.1.30) we obtain

$$\underline{U}_1 = \underline{K}_{11}^{-1}\left(\underline{\bar{F}}_1 - \underline{K}_{12}\underline{\bar{U}}_2\right). \tag{7.1.32}$$

Equation (7.1.31) gives the unknown forces which correspond to prescribed displacements. Such forces are called reactions.

The force method is a little more complicated and will not be developed in the present notes.

It will be shown now how the methods introduced in chapter 4 can be used to generate the discrete model.

7.2 Application of the Potential Energy Method

The generating displacements and displacement derivatives at a certain number of points of the generating domain called nodes or nodal points are taken for generalized displacements. Such displacements and displacement derivatives will be termed modal displacements along the present notes.

Transformation T is introduced by expressing the generating strains and displacements in terms of the generalized strains and displacements. We have then (see equations (4.7.21) and (4.7.26))

$$\left.\begin{aligned} \underline{e} &= \underline{\chi}^e \underline{e}^e \qquad (7.2.1)\\ \underline{u} &= \underline{\varphi}^e \underline{u}^e \qquad (7.2.2) \end{aligned}\right\} \text{within } \Delta^e .$$

According to (4.7.24) and (4.7.32) the expressions of $\underline{H}^e$ and $\underline{p}^e$ are

$$\underline{H}^e = \int_{\Delta^e} \underline{\chi}^{eT} \underline{H}\, \underline{\chi}^e \, d\Delta \tag{7.2.3}$$

$$\underline{p}^e = \int_{\Delta^e} \underline{\varphi}^{eT} \underline{\bar{f}} \, d\Delta + \int_{B^e \cap B_1} \underline{\varphi}^{eT} \underline{\bar{p}} \, dB + \int_{B^e \cap \Gamma} \underline{\varphi}^{eT} \underline{\bar{h}} \, d\Gamma . \tag{7.2.4}$$

An important point is that the expressions for the strains and displacements are in principle independent of each other, i.e. they have not to be related through the generating strain-displacement equations (4.2.2). In other words, compatibility can be violated not only across the element boundaries but also within the elements. On the other hand, as strains and displacements are mutually independent, it is not necessary to use the complicated expressions which often are termed consistent in the computation of the generalized forces associated to each node. Convergence can be obtained even if some easy rule for determining such forces is used, as the one which consists in ascribing to each node an equal fraction of the total external force acting in the element.

The generalized strain-displacement equations can be established in several ways.

A first way can be followed if the generating strains given by (7.2.1) are integrable, i.e. if there exists a generating displacement field related to such strains by the generating strain-displacement equations (4.2.2). It is easy then to express the nodal displacements in terms of the generalized strains and then, the generalized strains in terms of the nodal displacements. The resulting expressions are taken for generalized strain-displacement equations.

This is what happens in what we call the first modality in which the fields allowed within each element are defined through the displacements and the interaction between elements is defined through compatibility conditions.

A set of polynomial expressions are given then for the displacements and the coefficients of such expressions may be taken for generalized strains [39] . We have thus

$$\underset{\sim}{u}^e = \underset{\sim}{\Lambda}^e \underset{\sim}{e}^e \tag{7.2.5}$$

where the elements of $\underset{\sim}{\Lambda}^e$ are functions of the co-ordinates.

It must be observed that, if the coefficients of

the displacement expansions are taken for generalized strains, such generalized strains can be different from zero even if there is no deformation, i.e. if the body is given a rigid body displacement. Such inconvenient however is not serious and will not affect our conclusions.

Assuming that the initial strains vanish, the generating strains are given by

$$\underline{e} = \underline{D}\,\underline{\Lambda}^{e}\,\underline{e}^{e}. \tag{7.2.6}$$

Comparing (7.2.1) and (7.2.6) there results

$$\underline{\chi}^{e} = \underline{D}\,\underline{\Lambda}^{e}. \tag{7.2.7}$$

By successively particularizing equation (7.2. 5) for the different modal points, we obtain

$$\underline{u}^{e} = \underline{M}^{e}\underline{e}^{e} \tag{7.2.8}$$

and if the number of generalized strains equals the number of generalized displacements and $\underline{M}^{e}$ is non-singular, we obtain further

$$\underline{e}^{e} = \underline{M}^{e^{-1}}\underline{u}^{e}. \tag{7.2.9}$$

This classical procedure cannot be used sometimes, however, either because we have too many generalized strains (the displacement expansions contain too many coefficients, $\underline{M}^{e}$ is not a square matrix, and (7.2.9) has no meaning) or because the strains are not integrable (no continuous generating displacement field is connected to such strains through the generating strain-displacement equations).

In the first case, several general strain fields correspond to a given field of generalized displacements, and the total potential energy theorem can be used [40] for determining the field which is nearer to a generating equilibrated one.

In the second case we fall into what we call the second modality (hybrid elements), in which the fields allowed within each element are defined through the strains or stresses (°) and the interaction between elements is defined through compatibility conditions.

The analysis may start from equation (7.2.1). It is important, as we shall see, that all the fields allowed within the element equilibrate the same body forces. Such conditions can be fulfilled if such body forces are assumed to vanish. The allowed fields will be assumed thus to equilibrate vanishing body forces. This does not mean however that the prescribed body forces vanish. On the other hand, if it is really wished that the body forces within the element equal the prescribed body forces,

(°) In the second modality the allowed fields are traditionally defined through the stresses [2] . We observe however that no difference really exists in defining the fields through the stresses or through the strains, at least if the stress-strain equations are assumed linear.

it is always possible to add to the stress expansion a supplementary term which equilibrates the prescribed body forces [2] .

The complementary energy theorem can be used to determine, for each allowed strain field, the integrable field which equilibrates the same body forces (vanishing body forces, as assumed) and is nearer to it. Let us see how.

Assume first that we wish to determine the field which equilibrates vanishing body forces on a domain Δ^e and respects the displacement boundary conditions

$$\underline{u} = \overline{\underline{u}} \quad \text{on} \quad B^e. \tag{7 2.10}$$

The total complementary energy theorem states that such field is the one which minimizes the functional

$$T^{*e} = U^{*e} - \int_{\Delta^e} \underline{s}^T \underline{\bar{e}}^0 \, d\Delta - \int_{B^e} (\underline{N}\,\underline{s})^T \overline{\underline{u}} \, dB \tag{7.2.11}$$

in the set E_0 of the fields which equilibrate vanishing body forces.

As the allowed fields also equilibrate vanishing body forces, the set of the allowed fields is a subset of E_0. The allowed field which minimizes T^{*e} is thus, by virtue of the total complementary energy theorem (linear case) the one which is nearer to the compatible field which solves the problem.

Now by virtue of (7.2.1)

$$U^{*e} = \frac{1}{2}\int_{\Delta^e} \underline{s}^T \underline{H}^{-1} \underline{s}\, d\Delta = \frac{1}{2}\int_{\Delta^e} \underline{e}^T \underline{H}\, \underline{e}\, d\Delta =$$

(7.2.12)

$$= \frac{1}{2}\underline{e}^{eT}\int_{\Delta^e} \underline{\chi}^{eT} \underline{H}\, \underline{\chi}^e\, d\Delta\, \underline{e}^e = \frac{1}{2}\underline{e}^{eT}\underline{H}^e \underline{e}^e .$$

Introducing (7.2.2) into (7.2.11), and considering (7.2.12), we obtain

$$T^{*e} = \frac{1}{2}\underline{e}^{eT}\underline{H}^e\underline{e}^e - \underline{e}^{eT}\int_{\Delta^e}(\underline{H}\,\underline{\chi}^e)^T\underline{\bar{e}}^0\, d\Delta - \underline{e}^{eT}\int_{B^e}(\underline{N}\,\underline{H}\,\underline{\chi}^e)^T\underline{\varphi}^e\, dB =$$

(7.2.13)

$$= \frac{1}{2}\underline{e}^{eT}\underline{H}^e\underline{e}^e - \underline{e}^{eT}\underline{L}^e\underline{u}^e - \underline{e}^{eT}\int_{\Delta^e}(\underline{H}\,\underline{\chi}^e)^T\underline{\bar{e}}^0\, d\Delta$$

where the generalized displacements $\underline{u}^e$ are assumed prescribed. Matrix $\underline{L}^e$ is given by

(7.2.14) $$\underline{L}^e = \int_{B^e}(\underline{N}\,\underline{H}\,\underline{\chi}^e)^T\underline{\varphi}^e\, dB .$$

The first variation of T^{*e} is

(7.2.15) $$\delta T^{*e} = \delta\underline{e}^{eT}\left(\underline{H}^e\underline{e}^e - \underline{L}^e\underline{u}^e - \int_{\Delta^e}(\underline{H}\,\underline{\chi}^e)^T\underline{\bar{e}}^0\, d\Delta\right).$$

The stationarity condition is thus

(7.2.16) $$\underline{H}^e\underline{e}^e = \underline{L}^e\underline{u}^e + \int_{\Delta^e}(\underline{H}\,\underline{\chi}^e)^T\underline{\bar{e}}^0\, d\Delta$$

and if $\underline{H}^e$ is non-singular the following generalized strain-displacement equations result

$$\underline{e}^e = \underline{H}^{e^{-1}} \underline{L}^e \underline{u}^e + \underline{e}^{oe} \qquad (7.2.17)$$

where

$$e^{oe} = H^{e^{-1}} \int_{\Delta^e} (\underline{H}\,\underline{\chi}^e)^T \,\bar{\underline{e}}^0 \, d\Delta \, . \qquad (7.2.18)$$

Equation (7.2.17) coincides with the one established by Pian [2] but Pian's reasonment is different.

The interaction between elements is defined by compatibility conditions of the form (7.1.10), both in the first and second modalities, if the displacement discontinuities are assumed to vanish on Γ.

Different decisions can be made about which generating kinematic magnitudes are to be kept continuous at nodal points, i.e. about which magnitudes are to be taken for generalized displacements. The nodal values of the generating displacements are usually taken for generalized displacements but derivatives of such displacements may also be used. The important principle which must be kept in mind is that operator A_n must be bounded, i.e. that the norm, and thus the strain energy, of the approximation of any generating field with bounded energy must also be bounded. The discrete model will be called admissible if this condition is fulfilled.

Another principle which must be respected in order that convergence can be obtained is that compatibility must be preserved in the generalized-generating sense, which means that the T_n -image of a generating compatible field must be a generalized comaptible field.

This last principle will hold if

(i) The fulfillment of the generating strain-displacement equations by any compatible generating field implies the fulfillment of the generalized strain-displacement equations by the corresponding generalized field.

(ii) The nodal displacements of the T_n -image of any generating field are assumed equal to the nodal displacements of the generating field itself.

The fulfillment of condition (i) in the first modality is more or less evident. In what concerns the second modality, the proof is not difficult too.

Assume indeed

(7.2.19) $$\underline{e} = \underline{D}\,\underline{u} + \underline{e}^0 .$$

Introducing (7.2.1) and (7.2.2), there results

(7.2.20) $$\underline{\chi}^e \underline{e}^e = \underline{D}\,\underline{\varphi}^e \underline{u}^e + \underline{e}^0 .$$

Multiplying at left by $\underline{\chi}^{eT}\underline{H}$ and integrating on Δ^e, we obtain

$$\int_{\Delta^e} \underline{\chi}^{eT}\underline{H}\,\underline{\chi}^e d\Delta\, \underline{e}^e = \int_{\Delta^e} (\underline{H}\,\underline{\chi}^e)^T (\underline{D}\,\underline{\varphi}^e)\, d\Delta\, \underline{u}^e + \int_{\Delta^e} \underline{\chi}^{eT}\underline{H}\,\underline{e}^0 d\Delta$$

(7.2.21)

or by virtue of (7.2.3),

$$\underline{H}\,\underline{e}^e = \int_{\Delta^e} (\underline{H}\,\underline{\chi}^e)^T (\underline{D}\,\underline{\varphi}^e)\, d\Delta\, \underline{u}^e + \int_{\Delta^e} \underline{\chi}^{eT}\, \underline{H}\,\underline{e}^0\, d\Delta\,. \tag{7.2.22}$$

Considering the work principle, we obtain

$$\int_{\Delta^e} (\underline{H}\,\underline{\chi}^e)^T (\underline{D}\,\underline{\varphi}^e)\, d\Delta = \int_{\Delta^e} \left[\underline{E}\,(\underline{H}\,\underline{\chi}^e)\right]^T \underline{\varphi}^e\, d\Delta + \int_{B^e} (\underline{N}\,\underline{H}\,\underline{\chi}^e)^T \underline{\varphi}^e\, dB. \tag{7.2.23}$$

But, as the stresses allowed within the element equilibrate zero body forces

$$\underline{E}\,(\underline{H}\,\underline{\chi}^e) = \underline{0} \tag{7.2.24}$$

and therefore by virtue of (7.2.14),

$$\int_{\Delta^e} (\underline{H}\,\underline{\chi}^e)^T (\underline{D}\,\underline{\varphi}^e)\, d\Delta = \underline{L}^e. \tag{7.2.25}$$

Introducing into (7.2.22) there results (7.2.16) and this means that the generalized strain-displacement equations are respected.

In what concerns condition (ii), it suffices to remark that if this condition is satisfied, the generating compatibility equation (4.2.11) (with $\underline{\bar{g}} = \underline{0}$) implies the corresponding generalized equations (7.1.10).

Before proceeding further it is important to recognize that the first and second modalities are distinct, although they are both connected to the first generation method. More precisely, the first modality is not always a particular case of the second.

Indeed, while in the second modality the allowed fields were supposed to equilibrate vanishing body forces, they may equilibrate arbitrary body forces in the first modality.

However, we shall see that for the sake of convergence it is convenient that, even in the first modality the allowed fields equilibrate vanishing body forces. And then, if such practice is followed, the first modality really becomes a particular case of the second, characterized by the allowed strains being integrable.

7.3 Application of the Complementary Energy Method

If the discrete model is established according to the second generation method (third modality) traction resultants and moments, along each part of the boundary where the element is to be joined to a neighboring element, are to be taken for generalized forces.

The set of the allowed fields is defined by expressing the generating stresses and tractions in terms of the generalized stresses and forces. We have then (see equations (4.7.21) and (4.7.26))

$$\left.\begin{array}{l} \underline{s} = \underline{\psi}^e \underline{s}^e \\ \underline{p} = \underline{\gamma}^e \underline{f}^e \end{array}\right\} \text{within } \Delta^e. \qquad \begin{array}{r} (7.3.1) \\ (7.3.2) \end{array}$$

Expressions (7.3.1) and (7.3.2) are in principle quite independent of each other, i.e., they have not to be related through the generating traction-stress equations (4.2.4).

The generalized equilibrium equations can be established in several ways.

A first way [3] can be followed if the stresses and tractions are related by equations (4.2.4). Then indeed

$$\underline{p} = \underline{N}\, \underline{\psi}^e \underline{s}^e. \tag{7.3.3}$$

With the help of (7.3.3), we can compute traction resultants and moments and thus establish the linear relation (7.1.4) between $\underline{p}^e$ and $\underline{s}^e$.

Its tractions and stresses are not connected by (4.2.4), then the total complementary and the total potential energy theorems can be used for deriving the generalized traction-stress equations. The ways in which they may be used can be established if the analogy with the first method is considered.

According to (4.7.24) the stiffness matrix of the element si given by

$$\underline{H}^e = \left[\int_{\Delta^e} \underline{\psi}^{e^T} \underline{H}^{-1} \underline{\psi}^e \, d\Delta\right]^{-1} \tag{7.3.4}$$

and the generalized initial strains and prescribed displacements (see equations (4.7.31) and (4.7.32))

(7.3.5) $$\underline{e}^{oe} = \int_{\Delta^e} \underline{\psi}^{e^T} \underline{e}^o \, d\Delta$$

(7.3.6) $$\underline{\bar{u}}^e = \int_{\Delta^e} \underline{\gamma}^{e^T} \underline{\bar{u}} \, d\Delta \, .$$

The interaction between elements is defined by equations of the form (7.1.10), which express the equilibrium between the generalized tractions corresponding to contacting elements.

In order that equilibrium be preserved in the generalized-generating sense, the allowed field corresponding to a given generating field is defined to be the one which presents the same stress resultants and moments through the generalized equilibrium equations (7.1.1).

The discussion of completeness will be based on such definitions.

7.4 Completeness Criterion

The discussion of convergence in section 3.6 was based on the concept of completeness. A sequence of subsets C'_n of the subspaces X'_n is called complete with respect to another subset S of C, if, given a positive arbitrarily small number ε, it is possible to determine a positive integer N such that for any $n > N$, at least one element of C'_n exists within the neighborhood ε of each element belonging to S.

In the present case, the subsets C'_n contain the set of allowed fields corresponding to the different subdivisions of the domain into finite elements. Such allowed fields are assumed to verify the compatibility conditions (7.1.10), equilibrium conditions (7.1.13)). Subset C contains the set of the generating compatible (equilibrated fields).

It will be proved that the elements of C'_n which approximate the elements of a certain subset S of C are the T_n -images of such elements. Linearity will be assumed.

The distance between an arbitrary element $f \in C$ and its T_n -image f', can be estimated by considering the distance of both fields to an auxiliary field.

By virtue of the triangular inequality, we have indeed

(7.4.1) $$d(f, f') \leq d(f, f_t) + d(f_t, f').$$

Let S be the set of all the fields whose stresses (and strains) have bounded first order derivatives within each subdomain resulting from the subdivision of Δ .

Within a general subdomain, Δ^e, an arbitrary component of stress s_i of f can be expressed by

(7.4.2) $$s_i = s_i(0) + s_{i,j}(0_i)(x_j - x_j^0)$$

where 0 and 0_i are points of Δ^e . 0_i depends on the co-ordinates of the point where s_i is to be determined.

Let us consider now, within Δ^e , the stress field f_t^e with components

(7.4.3) $$s_{t_i} = s_i(0)$$

which we call tangent field to f at 0 . The auxiliary field f_t is nothing else than the piecewise fiele which coincides with f_t^e within each subdomain.

Let V_1^e be an upper bound of the first derivatives of the stresses within Δ^e . Equation (7.4.2) yields

(7.4.4) $$|s_i - s_{t_i}| < n_d V_1^e \ell^e$$

where ℓ^e is the maximum diameter of the element and n_d is the total number of the first derivatives.

Denoting by $\underline{s}$ and $\underline{s}_t$ the vectors of the stresses corresponding to f and f_t , and assuming linearity we have

$$\int_{\Delta^e} (\underline{s} - \underline{s}_t)^T H^{-1} (\underline{s} - \underline{s}_t) \, d\Delta < n_d^2 n_s h (V_1^e \ell^e)^2 \Delta^e \qquad (7.4.5)$$

where n_s denotes the number of stress components and h is an upper bound to the elements of matrix $\underline{H}^{-1}$.

The distance between f and f_t is therefore

$$d(f, f_t) = \sqrt{2 U_{f-f_t}} = \sqrt{2 \sum_e U_{f-f_t}^e} = \sqrt{n_d^2 n_s h \sum_e (V_1^e \ell^e)^2 \Delta^e} . \qquad (7.4.6)$$

If V_1 and ℓ denote the maximum values of V_1^e and ℓ_1^e in the whole set of subdomains Δ^e, we can simply write

$$d(f, f_t) \leq n_d \sqrt{n_s h \Delta} \cdot V_1 \ell \qquad (7.4.7)$$

which proves that the distance between f and f_t is of the order of ℓ :

$$d(f, f_t) = O(\ell) . \qquad (7.4.8)$$

We assume now that the tangent field is an allowed field. Such condition will be called completeness criterion for reasons which will become clear in the sequel.

If the completeness criterion is fulfilled, estimating the order of magnitude of the distance between f and f' becomes an easy job.

Indeed as f_t is an allowed field, i.e. as f_t belongs to the range X'_n of T_n, there results

(7.4.9) $$T_n(f_t) = f_t$$

and, as

(7.4.10) $$f' = T_n(f)$$

there follows

(7.4.11) $$f' - f_t = T_n(f - f_t).$$

Let now $\|T_n\|$ be the norm of the operator T_n which is assumed bounded. We have

(7.4.12) $$\|f' - f_t\| \leq \|T_n\| \cdot \|f - f_t\|$$

or

(7.4.13) $$d(f', f_t) \leq \|T_n\| \, d(f, f_t).$$

Introducing (7.4.13) in (7.4.1) there results

(7.4.13) $$d(f, f') \leq (1 + \|T_n\|) \, d(f, f_t)$$

and, therefore, by virtue of (7.4.8)

(7.4.14) $$d(f, f') = O(\ell).$$

This means that the distance between f and f' tends to zero with ℓ. Completeness is thus verified.

If the allowed fields are defined through the stresses as in the second and third modalities, the completeness criterion will be satisfied whenever the expression for each stress component contains an arbitrary constant term and the remaining terms can vanish regardless of the values taken by such constant.

The situation is a little more complicated if the allowed fields are defined through the displacements (first modality) in which case the expressions must be such that arbitrary rigid body displacements are allowed if the element is assumed free in the space.

Then, in case of two and three-dimensional elasticity, the completeness criterion will be satisfied if the expression for each displacement component contains an arbitrary constant term and all the linear terms affected by arbitrary constants. In the case of plates and if Kirchhoff's assumption is assumed to hold, i. e. if the transverse shear deformability is neglected, the expression for the transverse displacement must contain an arbitrary constant term and all the linear quadratic terms affected by arbitrary constants.
In either case the remaining terms must be able to vanish regardless of the values taken by the arbitrary constants.

The situation can be much more complicated in the case of shells (curved elements) since the system of coordinates ceases to be cartesian [41, 42] . A more general completeness criterion becomes necessary then: the completeness criterion may still require that the tangent field be an allowed field but the tangent field cannot be polynomial any more [20] .

7.5 Convergence Analysis

If compatibility (equilibrium) is preserved in the generalized-generating sense, the first convergence theorem can be used (first situation). Thus convergence will be obtained if completeness is achieved.

If compatibility (equilibrium) is not preserved in the generalized-generating sense, then the second convergence theorem must be used. Therefore, besides the completeness being achieved a supplementary condition is to be fulfilled: the sequence of approximate solution must converge to compatible (equilibrated) field. This condition is the one which is going to be discussed now.

The projection theorem states that the supplementary condition will be respected if the projection of the approximate solution in the set of compatible (equilibrated) fields, i.e. the compatible (equilibrated) field which equilibrates the same external forces (compatibilizes the same incompatibilities) as the approximate solution belongs to the set S , i.e.

to the set of the fields with bounded first derivatives of the stresses.

Assume first that the generating continuous models are such that operators $\underline{D}$ and $\underline{E}$ are of the first order as they are always if the macro-rotations or if the transverse shear deformability is not neglected.

Let us see first that in that case the first derivatives of the stresses remain bounded within each subdomain if the corresponding body force density components and the first derivatives of the corresponding initial strains remain also bounded within each subdomain as the size decreases indefinitely.

Consider Navier's equations (see section 4.2)

$$\left(\underline{E}\,\underline{H}\,\underline{D}\right)\underline{u} = \underline{f} \tag{7.5.1}$$

(7.5.1) shows that the components of $\underline{f}$ will be continuous and bounded if the derivatives of the displacements involved in $\left(\underline{E}\;\underline{H}\;\underline{D}\right)$ are continuous and bounded. On the other hand, a known theorem can be used to conclude that, reciprocally, the displacement derivatives with the same order as those involved in $\left(\underline{E}\;\underline{H}\;\underline{D}\right)$ will be continuous and bounded if the components of $\underline{f}$ are continuous and bounded.

The theorem which is known to be valid for linear elliptic equations of the form

(7.5.2) $$\alpha_{ij} u_{,ij} + \beta_i u_{,i} + j u = f$$

states that the partial derivatives of order up to $m+2$ of u satisfy a Hölder condition with exponent $\alpha\,(0<\alpha<1)$ in every bounded subdomain with closure in a closed domain $\bar{\Delta}^e$, whenever all the derivatives of order up to m of f, and the coefficients α_{ij}, β_i and j satisfy the same Hölder condition in $\bar{\Delta}^e$.

Assuming the theorem valid for elliptic equations of the form (7.5.1) we conclude that the displacements and their first and second derivatives will be continuous and bounded within Δ^e if the components of $\underline{f}$ are also continuous and bounded within $\bar{\Delta}^e$. Now,

(7.5.3) $$\underline{s} = \underline{H}\,\underline{e} = \underline{H}\,\underline{D}\,\underline{u} + \underline{H}\,\underline{e}^0$$

and thus

(7.5.4) $$\underline{s}_{,i} = (\underline{H}\,\underline{D}\,\underline{u})_{,i} + \underline{H}\,\underline{e}^0_{,i}$$

The first derivatives of the stresses will be bounded if the second derivatives of the displacements and the first derivatives of the initial strains are bounded or, considering what was concluded above if the body force density components and the first derivatives of the initial strains are continuous and bounded.

Assuming that the prescribed initial strains (body forces) have continuous and bounded first derivatives (density components), the initial strains corresponding to the projections are continuous and bounded. This means that the first derivatives of the stresses corresponding to the projections will be bounded if the body force density components (first derivatives of the initial strains) remain bounded and continuous within the elements as their size decreases indefinitely. Such is the condition for the projections belonging to **S** .

As the projection equilibrates (compatibilizes) the same external forces (incompatibilities) as the approximate solution, there results that in case the potential (complementary) energy method is employed, convergence will be obtained if the completeness criterion is satisfied and if the allowed fields (and thus the approximate solution) equilibrate (compatibilize) bounded and continuous body force density components (initial strains with bounded first derivatives) within each element. This will namely happen if allowed fields are used which equilibrate (compatibilize) vanishing body forces (initial strains).

Operators $\underline{\mathbf{D}}$ and $\underline{\mathbf{E}}$ cease to be of first order in the simplified theories which results from the corresponding exact theories by making the micro-rotations equal to the macro-rotations, or by neglecting the transverse shear deformability. The conclusions of the present discussion may be extended to such cases, as it is going to be demonstrated.

It must not be forgotten that, then, the body forces (initial strains) are not only the forces distributed within the subdomain but also the body moments (initial curvature). Convergence will be ensured only if both the body force and moment densities (initial strains and curvatures) remain continuous and bounded within the element as the size decreases indefinitely. Such condition is easily fulfilled if the allowed fields are defined through the stresses or strains, as in the second modality (hybrid elements) in which convergence will certainly be obtained if the allowed stress distributions equilibrate vanishing body forces and moments, as they were indeed assumed to do. The same happens in the third modality if the allowed fields are introduced through the displacements because then they compatibilize vanishing initial strains.

The condition is not so easily fulfilled in the first modality if the allowed fields are introduced through the displacements.

Let us examine indeed what happens for instance in the case of plates.

Combining equations (6.4.7-8), (6.11.3-5) and (6.11.9-12-13) we obtain the following equilibrium equations in terms of displacements

$$2\bar{\nu}\left(U_{3,11} + U_{3,22} + \Theta_{2,1} - \Theta_{1,2}\right) + F_3 = 0 \tag{7.5.5}$$

$$2\bar{\psi}\left(\Theta_{1,11} + \frac{\Theta_{1,22} + \Theta_{2,12}}{2}\right) + \bar{\varphi}\left(\Theta_{1,11} + \Theta_{2,21}\right) +$$

$$+ \bar{\chi}\left(\Theta_{1,22} - \Theta_{2,12}\right) + 2\bar{\nu}\left(U_{3,2} - \Theta_1\right) + G_1 = 0 \tag{7.5.6}$$

$$2\bar{\psi}\left(\Theta_{2,22}+\frac{\Theta_{2,11}+\Theta_{1,21}}{2}\right)+\bar{\varphi}\left(\Theta_{2,22}+\Theta_{1,12}\right)+$$

$$+\bar{\chi}\left(\Theta_{2,11}-\Theta_{1,21}\right)-2\bar{\vartheta}\left(U_{3,1}+\Theta_{2}\right)+G_{2}=0 \qquad (7.5.7)$$

which correspond to equation (7.5.1).

Equations (7.5.5-7) show that the body force and moment densities remain continuous and bounded within each element whenever the derivatives of order up to 2 of the transverse displacement and rotations remain also continuous and bounded.

Convergence will be obtained thus if completeness is achieved and the derivatives of order up to two of the transverse displacement and of the rotations are continuous and bounded.

Neglecting the transverse shear deformability is equivalent to making $\bar{\vartheta}=\infty$, case in which equations (7.5.5--7) cannot be used for the discussion of which derivatives must be kept continuous and bounded in order that convergence can be obtained.

Such difficulty may be removed however if the second convergence theorem respecting the second situation is applied (see section 3.6). Such theorem considers the convergence towards the same limit of a sequence of generating solutions and a sequence of approximate solutions in the general case of operator T_n not being conforming, i.e. in the present case of compatibility (equilibrium) being violated.

The exact generating solution corresponding to $\bar{\nu}=\infty$, can indeed be considered as the limit of a sequence of exact generating solutions corresponding to increasing values of $\bar{\nu}$ $(\bar{\nu}^1, \bar{\nu}^2, \ldots)$. Each set will thus obtain all the generating fields with stresses and strains related by elastic constants assuming the prescribed values, except $\bar{\nu}$ which takes the value $\bar{\nu}_n$. The subsets S_n contain the fields whose strains have bounded first derivatives.

On the other hand, the sequence of sets X'_n corresponds to the successive subdivisions of the domain into subdomains with decreasing size, the allowed fields within each subdomain being assumed such that the rotations are first derivatives of the transverse displacement. As it is also assumed that every state of constant curvature is allowed within each element, the distance between each field in S_n and its T_n - image in X'_n tends to zero as n tends to infinity, i.e. completeness is achieved.

The theorem states that convergence will be obtained if the successive projections of the successive approximate solutions in the successive sets of compatible (equilibrated) fields belong to the sets S_n and therefore if the derivatives of order up to two of the displacements remain bounded within each element as n tends to infinity, and as the rotations are first derivatives of the transverse displacement if the derivatives of order up to three of the transverse displacement remain bounded within each element as the size decreases indefinitely.

The supplementary condition is thus satisfied if the derivatives of order up to three of the transverse displacement are bounded.

Now the boundedness of the first and second derivatives is always ensured because the first derivatives are proportional to rotations and the second derivatives represent strains which must be bounded since operator A_n is bounded. The same does not happen with the third derivatives however, and this explains why convergence is sometimes obtained for some patterns of elements and not for others (°).

This is namely what happens with the well-known plate element which Bazeley, Cheung, Irons and Zienkiewicz [7] introduced in 1965, for which numerical experience indicates that convergence is obtained with the pattern represented in fig. 7.1 but not with the one represented in fig. 7.2.

It is easy to demonstrate that the third deriva-

(°) We repeat once again that if compatibility is achieved across the element boundaries, convergence will always be obtained once completeness is achieved, independently of the pattern being used.

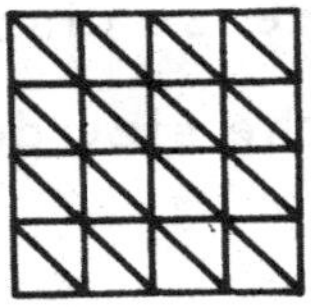

Fig. 7.1

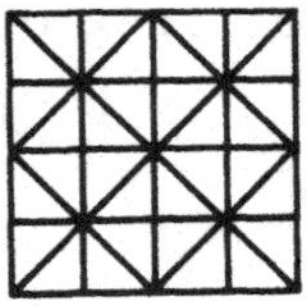

Fig. 7.2

tives of the transverse displacement cannot become unbounded in the first case because there is only one kind of nodes, i.e. all the nodes are in the same conditions. The nodal values of the transverse displacement and their first derivatives have thus a smooth discrete variation from node to node if the same happens to the generalized external forces ascribed to each node, i.e. if the external forces distribute continuously on Δ . The third derivatives of the transverse displacement must then remain bounded on each edge of every element. As there are only four derivatives of third order and these are related between themselves and to derivatives of lower order (which are all bounded) by a condition which reflects the fact that the number of degrees of freedom within each element is 9 and not 10, as it would be if all the coefficients of the third degree polynomial were arbitrary, there follows that the third derivatives of the transverse displacement are all bounded and convergence is thus obtained.

7.6 The Mixed Method

In the preceding sections it has been shown that the three first finite element modalities are generally related to one of the two generation methods described in chap. 5. The so-called mixed models [4, 42] are however an exception. They are indeed related to a mixed method which will be considered now.

The mixed method can be applied if matrices $\underline{D}$, $\underline{E}$, $\underline{N}$ and $\underline{H}$ of the generating model have the form

$$\underline{D} = \begin{bmatrix} \underline{D}_1 & \underline{I} \\ \underline{O} & \underline{D}_2 \end{bmatrix}; \quad \underline{E} = \begin{bmatrix} \underline{E}_1 & \underline{O} \\ \underline{I} & \underline{E}_2 \end{bmatrix}; \quad \underline{N} = \begin{bmatrix} \underline{N}_1 & \underline{O} \\ \underline{O} & \underline{N}_2 \end{bmatrix}; \quad \underline{H} = \begin{bmatrix} \underline{H}_1 & \underline{O} \\ \underline{O} & \underline{H}_2 \end{bmatrix} \qquad (7.6.1-4)$$

in which $\underline{I}$ is a unit matrix. We remark that $\underline{H}_1$ and $\underline{H}_2$ must be positive definite in order that $\underline{H}$ be positive definite.

Considering vectors $\underline{s}$, $\underline{e}$, $\underline{u}$ and $\underline{f}$ partitioned in conformity with $\underline{E}$, $\underline{D}$, $\underline{N}$ and $\underline{H}$ the generating equations can take the form

$$\begin{cases} \underline{e}_1 = \underline{e}_1^0 + \underline{D}_1 \underline{u}_1 + \underline{u}_2 \\ \underline{e}_2 = \underline{e}_2^0 + \underline{D}_2 \underline{u}_2 \end{cases}; \quad \begin{cases} \underline{E}_1 \underline{s}_1 = \underline{f}_1 \\ \underline{s}_1 + \underline{E}_2 \underline{s}_2 = \underline{f}_2 \end{cases}; \quad \begin{cases} \underline{N}_1 \underline{s}_1 = \underline{p}_1 \\ \underline{N}_2 \underline{s}_2 = \underline{p}_2 \end{cases}; \quad \begin{cases} \underline{s}_1 = \underline{H}_1 \underline{e}_1 \\ \underline{s}_2 = \underline{H}_2 \underline{e}_2 . \end{cases}$$

(7.6.5 - 8)

Two new auxiliary structural generating models will be considered now. The first one involves magnitudes $\underline{s}_1, \underline{e}_1, \underline{u}_1, \underline{f}_1, \underline{p}_1, \ldots$, interrelated by equations

(7.6.9) $$\underline{E}_1 \underline{s}_1 = \underline{f}_1$$

(7.6.10) $$\underline{e}_1 = \underline{d}_1^0 + \underline{D}_1 \underline{u}_1$$

(7.6.11) $$\underline{s}_1 = \underline{H}_1 \underline{e}_1$$

(7.6.12) $$\underline{N}_1 \underline{s}_1 = \underline{p}_1$$

The second one involves magnitudes $\underline{s}_2, \underline{e}_2, \underline{u}_2, \underline{f}_2, \underline{p}_2, \ldots$, interrelated by

(7.6.13) $$\underline{E}_2 \underline{s}_2 = \underline{h}_2$$

(7.6.14) $$\underline{e}_2 = \underline{e}_2^0 + \underline{D}_2 \underline{u}_2$$

(7.6.15) $$\underline{s}_2 = \underline{H}_2 \underline{e}_2$$

(7.6.16) $$\underline{N}_2 \underline{s}_2 = \underline{p}_2 .$$

The initial strains of the first auxiliary model and the body forces of the second are respectively denoted by $\underline{d}_1^0$ and $\underline{h}_2$, in order that they can be distinguished from the corresponding initial strains and body forces of the initial model.

The auxiliary models will be true structural models only if the operators $\underline{D}_1, \underline{E}_1, \underline{N}_1$ and $\underline{D}_2, \underline{E}_2, \underline{N}_2$, are such that the work principle holds in both. This is so because operators $\underline{D}, \underline{E}, \underline{N}$ are such that

(7.6.17) $$\int_{\Delta^e} \underline{s}^T (\underline{D}\, \underline{u})\, d\Delta = \int_{\Delta^e} (\underline{E}\, \underline{s})^T \underline{u}\, d\Delta + \int_{B^e} (\underline{N}\, \underline{s})^T \underline{u}\, dB$$

Introducing (7.6.1-4) and particularizing for $\underline{\sigma}_2 = \underline{0}$ and $\underline{u}_2 = \underline{0}$ we obtain

$$\int_{\Delta^e} \underline{\sigma}_1^T (\underline{D}_1 \underline{u}_1)\, d\Delta = \int_{\Delta^e} (\underline{E}_1 \underline{\sigma}_1)^T \underline{u}_1\, d\Delta + \int_{B^e} (\underline{N}_1 \underline{\sigma}_1)^T \underline{u}_1\, dB \tag{7.6.18}$$

from which we obtain, by introducing (7.6.9-12),

$$\int_{\Delta^e} \underline{\sigma}_1^T \underline{e}_1\, d\Delta = \int_{\Delta^e} \underline{\sigma}_1^T \underline{d}_1^0\, d\Delta + \int_{\Delta^e} \underline{f}_1^T \underline{u}_1\, d\Delta + \int_{B^e} \underline{p}_1^T \underline{u}_1\, dB\, . \tag{7.6.19}$$

Particularizing for $\underline{\sigma}_1 = \underline{0}$ and $\underline{u}_2 = \underline{0}$ there results

$$\int_{\Delta^e} \underline{\sigma}_2^T \underline{e}_2\, d\Delta = \int_{\Delta^e} \underline{\sigma}_2^T \underline{e}_2^0\, d\Delta + \int_{\Delta^e} \underline{h}_2^T \underline{u}_2\, d\Delta + \int_{B^e} \underline{f}_2^T \underline{u}_2\, dB\, . \tag{7.6.20}$$

Equations (7.6.19) and (7.6.20) express the work principle for each of the auxiliary models.

The mixed method considers the initial structure as resulting from the association of two fictitious structures respectively governed by the equations (7.6.9 - 12) and (7.6.13-16). Such structures do not behave independently because the initial strains in the first depend on the displacements of the second and the body forces in the second depend on the stresses of the first. Indeed comparing (7.6.5) to (7.6.10) and (7.6.6) to (7.6.13), there results

$$\underline{d}_1^0 = \underline{e}_1^0 + \underline{u}_2 \tag{7.6.21}$$

(7.6.22) $$\underline{h}_2 = \underline{f}_2 - \underline{\eth}_1$$

$\underline{e}^0_1$ and $\underline{f}_2$ will be assumed to vanish in what follows.

Suppose now that both fictitious structures are discretized by the finite element technique. Assume first that the first generation method is applied to the first structure and the second generation method to the second structure.

Thus we start from

(7.6.23) $$\underline{e}_1 = \underline{\chi}^e_1\, \underline{e}^e_1$$

(7.6.24) $$\underline{u}_1 = \underline{\varphi}^e_1\, \underline{u}^e_1$$

(7.6.25) $$\underline{\eth}_2 = \underline{\psi}^e_1\, \underline{\eth}^e_2$$

(7.6.26) $$\underline{p}_2 = \underline{\gamma}^e_2\, \underline{p}^e_2 \,.$$

The generalized strain-displacement equations are obtained in the first structure by assuming $\underline{\varphi}^e_1$ and $\underline{\chi}^e_1$ to be such that the generating strain-displacement equations (7.6.10) are respected. Similarly the generalized equilibrium conditions result from assuming matrices $\underline{\psi}^e_2$ and $\underline{\gamma}^e_2$ to be such that the generating equilibrium equations (7.6.13) and (7.6.16) are satisfied.

Introducing (7.6.22) in (7.6.13) and making $\underline{f}_2 = \underline{0}$, there results

(7.6.27) $$\underline{E}_2\, \underline{\eth}_2 = -\underline{\eth}_1$$

Introducing (7.6.25) and (7.6.11) there results

$$\underline{e}_1 = - \underline{H}^{-1} \underline{E}_2 \underline{\psi}_2^e \underline{\triangle}_2^e . \tag{7.6.28}$$

Condition (7.6.22) will be fulfilled if the generalized stresses in the second structure are taken for generalized strains in the first and if

$$\underline{\chi}_1^e = - \underline{H}_1^{-1} \underline{E}_2 \underline{\psi}_2^e . \tag{7.6.29}$$

Introducing (7.6.21) in (7.6.10) and making $\underline{e}_1^0 = \underline{0}$, there results

$$\underline{e}_1 = \underline{u}_2 + \underline{D}_1 \underline{u}_1 . \tag{7.6.30}$$

Introducing (7.6.23) and (7.6.24) there results

$$\underline{u}_2 = \underline{\chi}_1^e \underline{e}_1^e - \underline{D}_1 \underline{\varphi}_1^e \underline{u}_1^e \tag{7.6.31}$$

or

$$\underline{u}_2 = - \underline{H}_1^{-1} \underline{E}_2 \underline{\psi}_2^e \underline{\triangle}_2^e - \underline{D}_1 \underline{\varphi}_1^e \underline{u}_1^e . \tag{7.6.32}$$

Condition (7.6.21) will be satisfied thus if equation (7.6.32) is also satisfied.

The approximate solution corresponding to the first fictitious structure can be obtained by minimizing the total potential energy

$$T_1 = U_1 - \sum_e \int_{\Delta^e} \underline{\bar{f}}_1^T \underline{u}_1 \, d\Delta - \int_{B_1} \underline{\bar{p}}_1^T \underline{u}_1 \, dB \tag{7.6.33}$$

in the class of the generalized compatible fields.

The first variation δT_1, is

$$\delta T_1 = \sum_e \int_{\Delta^e} \underline{s}_1 T \delta \underline{e}_1 \, d\Delta - \sum_e \int_{\Delta^e} \underline{\bar{F}}_1 \delta \underline{u}_1 \, d\Delta - \int_{B_1} \underline{\bar{p}}_1 \delta \underline{u}_1 \, dB . \tag{7.6.34}$$

But

$$\sum_e \int_{\Delta^e} \underline{\bar{F}}_1^T \delta \underline{u}_1 \, d\Delta + \int_{B_1} \underline{\bar{p}}^T \delta \underline{u}_1 \, dB = \sum_e \underline{\bar{F}}_1^{e^T} \delta \underline{u}_1^e \tag{7.6.35}$$

in which

$$\underline{\bar{F}}_1^e = \int_{\Delta^e} \underline{\varphi}_1^{e^T} \underline{\bar{F}}_1 \, d\Delta + \int_{B^e \cap B_1} \underline{\varphi}_1^{e^T} \underline{\bar{p}}_1 \, dB . \tag{7.6.36}$$

On the other hand, as $\underline{\varphi}_1^e$ and $\underline{\chi}_1^e$ were assumed such that the generating strain-displacement equations are satisfied.

$$\delta \underline{e}_1 = \delta \underline{d}_1^0 + \underline{D}_1 \delta \underline{u}_1 . \tag{7.6.37}$$

But as the total potential energy theorem refers to the first auxiliary model, $\underline{d}_1^0$ must be assumed to be prescribed so that

$$\delta \underline{e}_1 = \underline{D}_1 \delta \underline{u}_1 = \underline{D}_1 \underline{\varphi}_1^e \delta \underline{u}_1^e . \tag{7.6.38}$$

Thus introducing (7.6.28) and (7.6.38) we obtain

$$\int_{\Delta^e} \underline{s}_1^T \delta \underline{e}_1 \, d\Delta = - \underline{s}_2^{e^T} \int_{\Delta^e} \left(\underline{E}_2 \underline{\psi}_2^e\right)^T \left(\underline{D}_1 \underline{\varphi}_1^e\right) d\Delta \, \delta \underline{u}_1^e = - \underline{s}_2^{e^T} \underline{J}^e \delta \underline{u}_1^e \tag{7.6.39}$$

in which

$$\underline{J}^e = \int_{\Delta^e} (\underline{E}_2 \underline{\psi}_2^e)^T (\underline{D}_1 \underline{\varphi}_1^e)\, d\Delta . \tag{7.6.40}$$

The first variation δT_1 becomes thus

$$\delta T_1 = -\sum_e \underline{\Delta}_2^{eT} \underline{J}^e \delta \underline{u}_1^e + \sum_e \underline{\bar{F}}_1^e \delta \underline{u}_1^e = 0 . \tag{7.6.41}$$

Introducing now the generalized displacement discontinuity relations

$$\underline{u}_1^e = \underline{A}_1^e \underline{U}_1 \tag{7.6.42}$$

we obtain

$$\sum_e (\underline{J}^e \underline{A}_1^e)^T \underline{\Delta}_2^e = \sum_e (\underline{A}_1^e \underline{\bar{F}}_1^e) . \tag{7.6.43}$$

The approximate solution corresponding to the second fictitious structure can be obtained by minimizing the total complementary energy

$$T_2^* = U_2^* - \sum_e \left(\int_{\Delta^e} \underline{\Delta}_2^T \underline{\bar{e}}_2^0 \, d\Delta \right) - \int_{\underline{B}_2} \underline{p}_2^T \underline{\bar{u}}_2 \, dB \tag{7.6.44}$$

in the class of the generalized equilibrated fields.

As the generalized equilibrium conditions were assumed to result from the fulfillment of the generating equilibrium equations within each element, the Lagrange's multiplier method can be used and the stationarity conditions determined for functional

$$T_2^{**} = T_2^{*} + \sum_e \int_{\Delta^e} (\underline{h}_2 - \underline{E}_2 \underline{s}_2)^T \underline{u}_2 \, d\Delta \tag{7.6.45}$$

in which displacements $\underline{u}_2$ appear as multipliers.

The first variation is

$$\delta T_2^{**} = \sum_e \int_{\Delta^e} \delta \underline{s}_2^T \underline{e}_2 \, d\Delta - \sum_e \int_{\Delta^e} \delta \underline{s}_2^T \bar{\underline{e}}_2^0 \, d\Delta - \int_{B_2} \delta \underline{p}_2^T \bar{\underline{u}}_2 \, dB -$$
(7.6.46)
$$- \sum_e \int_{\Delta^e} (\underline{E}_2 \delta \underline{s}_2)^T \underline{u}_2 \, d\Delta + \sum_e \int_{\Delta^e} (\underline{h}_2 - \underline{E}_2 \underline{s}_2)^T \delta \underline{u}_2 \, d\Delta .$$

(7.6.46)

The variation of $\underline{h}_2$ was assumed to vanish because the total complementary energy theorem is considered in the frame of the second auxiliary model.

Introducing (7.6.25) and (7.6.26) there results

$$\sum_e \int \delta \underline{s}_2^T \bar{\underline{e}}_2^0 \, d\Delta + \int \delta \underline{p}_2^T \bar{\underline{u}}_2 \, dB = \sum_e \int \delta \underline{s}_2^{eT} \bar{\underline{d}}_2^e \tag{7.6.47}$$

in which

$$\bar{\underline{d}}_2^e = \int_{\Delta^e} \underline{\psi}_2^{eT} \bar{\underline{e}}_2^0 \, d\Delta + \int_{B^e \cap B_2} \underline{\gamma}_2^{eT} \bar{\underline{u}}_2 \, dB \tag{7.6.48}$$

On the other hand, introducing (7.6.25) and (7.6.32), we obtain

$$\int_{\Delta^e} (\underline{E}_2 \delta \underline{s}_2)^T \underline{u}_2 \, d\Delta = - \delta \underline{s}_2^{eT} \int_{\Delta^e} (\underline{E}_2 \underline{\psi}_2^e)^T (\underline{D}_1 \underline{\varphi}_1^e) \, d\Delta \, \underline{u}_1^e -$$
$$- \delta \underline{s}_2^{eT} \int_{\Delta^e} (\underline{E}_2 \underline{\psi}_2^e)^T \underline{H}_1^{-1} (\underline{E}_2 \underline{\psi}_2^e) \, d\Delta \, \underline{s}_2^e = - \delta \underline{s}_2^{eT} \underline{J}^e \underline{u}_1^e -$$
(7.6.49a)

$$-\delta \underline{\sigma}_2^{e^T} \int_{\Delta^e} (\underline{E}_2 \underline{\psi}_2^e)^T \underline{H}_1^{-1} (\underline{E}_2 \underline{\psi}_2^e) \, d\Delta . \qquad (7.6.49b)$$

Also

$$\int_{\Delta^e} \delta \underline{\sigma}_2^T \underline{e}_2 \, d\Delta = \delta \underline{\sigma}_2^e \int_{\Delta^e} \underline{\psi}_2^{e^T} \underline{H}_2^{-1} \underline{\psi}_2^e \, d\Delta \, \underline{\sigma}_2^e . \qquad (7.6.50)$$

Variation δT_2^{**} transforms thus into

$$\delta T_2^{**} = \sum_e \delta \underline{\sigma}_2^e (\underline{S}^e \underline{\sigma}_2^e - \underline{\bar{d}}_2^e + \underline{J}^e \underline{u}_1^e) + \sum_e \int_{\Delta^e} (\underline{h}_2 - \underline{E}_2 \underline{\sigma}_2)^T \delta \underline{u}_2 \, d\Delta \qquad (7.6.51)$$

where

$$\underline{S}^e = \int_{\Delta^e} \underline{\psi}_2^{e^T} \underline{H}_2^{-1} \underline{\psi}_2^e \, d\Delta + \int_{\Delta^e} (\underline{E}_2 \underline{\psi}_2^e)^T \underline{H}_2^{-1} (\underline{E}_2 \underline{\psi}_2^e) \, d\Delta . \qquad (7.6.52)$$

As the stress fields inside the different elements are interrelated by generalized equilibrium conditions, the stress variation vectors $\delta \underline{\sigma}_2^e$ cannot be all independent.

The equilibrium conditions can be however of different nature. Sometimes the stress vector $\underline{\sigma}_2$ is assumed to be continuous at the nodes. Sometimes tractions $\underline{p}_2$ are assumed to equilibrate at the middle point of each edge or face.

Assume for instance that the stress vector is continuous at the nodes and call $\underline{S}_2$ the vector of the nodal stresses. It is then possible to write

(7.6.53) $$\underline{\sigma}_2^e = \underline{A}_2^e \underline{S}_2$$

for each element. Equation (7.6.53) is similar to eq. (7.6.42) which was used for displacements $\underline{u}_1$.

Introducing (7.6.53) in (7.6.51), we obtain

$$\delta T^{**} = \delta \underline{S}_2^T \sum_e \left(\underline{A}_2^{eT} \underline{S}^e \underline{A}_2^e \underline{S}_2 - \underline{A}_2^{eT} \underline{\bar{d}}_2^e + \underline{A}_2^{eT} \underline{J}^e \underline{u}_1^e \right) + \sum_e \int_{\Delta^e} (\underline{h}_2 - \underline{E}_2 \underline{\sigma}_2)^T \delta \underline{u}_2 d\Delta .$$

(7.6.54)

The stationary conditions are thus the equilibrium equation (7.6.13) and, as $\delta \underline{S}_2$ is arbitrary, the equation

(7.6.55) $$\sum_e \left(\underline{A}_2^{eT} \underline{S}^e \underline{A}_2^e \right) \underline{S}_2 + \sum_e \left(\underline{A}_2^{eT} \underline{J}^e \underline{A}_1^e \right) \underline{U}_1 - \sum_e \left(\underline{A}_2^{eT} \underline{\bar{d}}_2^e \right).$$

Such equation must be combined with equation (7.6.43) which can be transformed into

(7.6.56) $$\sum_e \left(\underline{A}_2^{eT} \underline{G}^e \underline{A}_1^e \right)^T \underline{S}_2 = \sum_e \left(\underline{A}_1^{eT} \underline{\bar{F}}_1^e \right)$$

by introducing (7.6.53). These equations are identical to the ones presented by Connor [42] and are only formally different from those presented by Herrmann [4] .

Convergence will be obtained with the mixed model if the general convergence criteria are satisfied by both auxiliary models.

Completeness is thus a condition to be satisfied in both. It will be satisfied in the first if matrices $\underline{\chi}_1^e$ are such that stresses $\underline{\sigma}_1$ can take arbitrary constant values within each element. It will be satisfied in the second if matrices $\underline{\psi}_2^e$ are such that the same happens to stresses $\underline{\sigma}_2$.

However, $\underline{\chi}_1^e$ is connected to $\underline{\psi}_2^e$ through (7.6.29). This means that, if $\underline{E}_2$ is a first order operator, completeness will be achieved if stresses $\underline{\sigma}_2$ and their first derivatives can take arbitrary constant values within each element. Such is for instance the case if the allowed stresses $\underline{\sigma}_2$ are defined by linear expressions with arbitrary coefficients.

Convergence will be obtained if completeness is achieved and the generalized compatibility and equilibrium conditions are such that continuity of displacements $\underline{u}_1$ and equilibrium of stresses $\underline{\sigma}_2$ are satisfied across the element boundaries. We remark that no other supplementary compatibility or equilibrium condition is necessary, as the strain-displacement equations of the first model and the equilibrium equations of the second were assumed to be satisfied within each element.

Compatibility and/or equilibrium across the element boundaries are however not always necessary. Convergence will indeed still be achieved if $\underline{E}_1\underline{\sigma}_1$, i.e. the body forces equilibrated by stresses $\underline{\sigma}_1$, and $(\underline{\varepsilon}_2-\underline{D}_2\underline{u}_2)$, i.e. the initial strains compatibilized by strains $\underline{\varepsilon}_2$ and displacements $\underline{u}_2$, remain bounded within each element. Considering (7.6.27), this will happen if

$$\underline{E}_1\,\underline{E}_2\,\underline{\psi}_2^e = \underline{0} \tag{7.6.57}$$

and if stresses $\underset{\sim}{\mathfrak{s}}_1$ are integrable, i.e. if a field exists such
that

(7.6.58) $\underline{\psi}_2^e \, \underline{\mathfrak{s}}_2^e = \underline{H}_2 \, \underline{D}_2 \, \underline{u}_2$

for an arbitrary $\underline{\mathfrak{s}}_2^e$.

If strains $\underline{e}_1$ are neglected (transverse shear
strain neglected) stresses $\underset{\sim}{\mathfrak{s}}_1$ will not contribute for strain energy and it is no longer necessary that they can take arbitrary
values in order that completeness can be achieved. It is then
enough that stresses $\underset{\sim}{\mathfrak{s}}_2$ can take arbitrary constant values
within each element.

Let us apply now such conclusions to the mixed models which have been proposed for plates.

The condition for completeness is that the expressions for the moments present all the constant and linear
terms affected by arbitrary coefficients. Only the constant
terms need to be arbitrary however, if the transverse shear
strain is neglected.

Convergence will be obtained if besides completeness being satisfied, the allowed transverse displacement
is continuous across the element boundaries and the allowed
moments satisfy the equilibrium conditions on such boundaries.
This happens namely with triangular elements if the vertices
are taken for nodes and the displacements and moments vary
linearly within the elements (continuous moment and displacement fields).

If the continuity of the transverse displacement or the equilibrium of moments are violated across the element boundaries, convergence will be obtained if completeness is still maintained and the allowed moments satisfy the conditions (see equation (6.11.17))

$$M_{12,11} + \left(M_{22} - M_{11}\right)_{,12} - M_{21,22} = 0 \qquad (7.6.59)$$

(which is nothing else than equation (7.6.57)) and

$$\nabla^2\left(M_{12} - M_{21}\right) = 0 \qquad (7.6.60)$$

which corresponds to equation (7.6.58).

Convergence will thus still be obtained under the assumption of the transverse shear strain being neglected, if (see [42]) constant moment fields are used over such element and the only condition imposed for generalized equilibrium is the equality of the bending moments applied to adjacent edges.

Note - It is assumed in equations (7.6 59) and (7.6.60) that the moments are defined according to equation (6.9.2) (with $\lambda_{\bar{\alpha}} = 1$) and not according to

$$M_{\alpha\beta} = \int_{-t/2}^{t/2} \xi_3 \, \sigma_{\alpha\beta} \, d\xi_3$$

as it is more usual. If this last definition were used, equations (7.6.59) and (7.6.60) would take the forms

$$M_{11,11} + 2\,M_{12,22} + M_{22,22} = 0 \qquad (7.6.59')$$

$$\nabla^2 (M_{11} + M_{22}) = 0 \qquad (7.6.60')$$

under which they were introduced in

As it is well-known, equation (7.6.59') results from eliminating the shear forces from the equilibrium equations with vanishing body forces. Equation (7.6.60') is the integrability condition for the moments which may be obtained by combining the compatibility conditions for the curvatures with the moment-curvature equations and equation (7.6.59').

REFERENCES

[1] Turner M. J., Clough R. W., Martin H. C. and Topp L. J.: "Stiffness and Deflection Analysis of Complex Structures", J. Aeron. Sci. 23 (9), 1956.

[2] Pian T. H.: "Derivation of Element Stiffness Matrices by Assumed Stress Distributions", AIAA Jour. (7), July 1964.

[3] Veubeke B. F. de: "Upper and Lower Bounds in Matrix Structural Analysis", AGARDograph 72, 165, Pergamon Press, 1964.

[4] Herrmann L.: "A Bending Analysis of Plates", Proc. 1st Conference on Matrix Methods in Structural Mechanics, Wright-Patterson A. F. B., 1965.

[5] Clough R. W.: "The Finite Element Method in Structural Mechanics, Philosophy of the Finite Element Procedure", in "Stress Analysis" (Ed. by O. C. Zienkiewicz and G. S. Holister), Chap. 7, London, John Wiley and Sons Ltd. 1965.

[6] Papenfuss S. W.: "Lateral Plate Deflection by Stiffness Matrix Methods with Application to a Marquee" M. S. Thesis, Dept. of Civil Engineering, Un. of Washington, Seattle, December 1959.

[7] Bazeley G. R., Cheung Y. K., Irons B. M. and Zienkiewicz O. C.: "Triangular Elements in Plate Bending. Conforming and Non-conforming Solutions" Proc. of the 1st Conference on Matrix Methods in Structural Mechanics, Wright-Pat

terson A. F. B., Ohio 1965.

[8] Irons B. and Draper K.: "Inadequacy of Nodal Connections in a Stiffness Solution for Plate Bending", AIAA Journal 3 1965 (p. 965).

[9] Key S. W.: "A Convergence Investigation of the Direct Stiffness Method" Ph. D. Dissertation University of Washington 1966.

[10] Tong P. and Pian T. H. H.: "The Convergence of Finite Element Method in Solving Linear Elastic Problems" Int. J. Solid Structures, vol. 3, 1967 pp. 865-879.

[11] Mikhlin S. G.: "Variational Methods in Mathematical Physics", New York, Pergamon Press 1964.

[12] Johnson M. W. Jr. and McLay R. W.: "Convergence of the Finite Element Method in the Theory of Elasticity" J. Applied Mechanics, Paper n. 68 APM-2.

[13] Oliveira E. R. A.: "Theoretical Foundations of the Finite Element Method" Int. J. Solids Structures, vol. 4 No. 10, October 1968, pp. 929-952.

[14] Oliveira E. R. A.: "A Theory of Shells Involving Moments of Arbitrary Order", Lisboa LNEC, November 1967.

[15] Oliveira E. R. A.: "A Further Research on the Analysis of Shells Using Moments of Arbitrary Order", Lisboa LNEC, December 1967.

[16] Irons B. M., Zienkiewicz O. C. and Oliveira E. R. A.: "Comment on the Paper: Theoretical Foundations of the Finite Element Method", Int. J. Solids Structures, Volume 6, pp. 695-697.

[17] Oliveira E. R. A.: "Completeness and Convergence in the Finite Element Method", Proc. of the 2nd Conference on Matrix Methods in Structural Mechanics, Wright-Patterson AFB, Ohio 1968.

[18] Oliveira E. R. A.: "Introduçao à Teoria das Estruturas" Lisboa Ediçao do Autor, 1966.

[19] Oliveira E. R. A.: "The Convergence Theorems and Their Role in the Theory of Structures", IUTAM Symposium on High Speed Compûting of Elastic Structures, Liège 1970.

[20] Oden J. T.: "Finite Element of Nonlinear Continua", McGraw-Hill, New York 1958.

[21] Taylor A. E.: "Introduction to Functional Analysis" John Wiley, New York 1972.

[22] Lusternik L. A. and Sobolev V. J.: "Elements of Functional Analysis" Hindustan Publishing Corp. New Delhi, 1961.

[23] Veubeke B. F. de: "Displacements and Equilibrium Models in the Finite Element Method", in O. C. Zienkiewicz and G. S. Holister (editors), "Stress Analysis", J. Wiley, New York 1965.

[24] Reissner E.: "On a Variational Theorem in Elasticity" J. Math. Phys., 29, 1950 (p. 90).

[25] Kröner E. (Ed.): "Proceedings of the IUTAM Symposium on "Mechanics of Generalized Continua" Springer 1968.

[26] Cosserat E. and F.: "Théorie des Corps Déformables" A. Hermann & Sons, Patis 1909.

[27] Truesdell C. and Noll N.: "The Nonlinear Field Theories of Mechanics" in "Encyclopedia of Physics", vol. III/3, Springer-Verlag, N. York 1965.

[28] Toupin R. A.: "Theories of Elasticity with Couple-Stresses", Arch. Rat. Mech. Anal. 17, 85-112 (98), 1964.

[29] Mindlin R. D.: "Microstructure in Linear Elasticity" Arch. Rat. Mech. Anal. 16, 51-78 (98), 1964.

[30] Green A. E. and Rivlin R. S.: "Mulitpolar Continuum Mechanics", Arch. Rat. Mech. Anal., 17, 113-147, 1964.

[31] Eringen A. C. and Suhubi E. S.: "Nonlinear Theory of Simple Microelastic Solids", Int. J. Engng. Sc. 2 (2 and 4), 1964.

[32] Grioli G.: "Elasticità Asimmetrica", Annali di Matematica Pura ed Applicata, Bologna, Serie IV, Tomo L, 1960.

[33] Paria G.: "Constitutive Equations in Cosserat Elasticity", J. of Engineering Math., vol. 4 (1970) 203-208.

[34] Newell H. E.: "Vector Analysis" in Handbook of Engineering Mechanics, McGraw-Hill, 1962.

[35] Mindlin R. D. and Tierten H. F.: "Effects of Couple-Stresses in Linear Elasticity", Arch. Rat. Mech. Anal. 11 (5), 1962.

[36] Reissner E.: "Variational Considerations for Elastic Beams and Shells", J. of the Eng. Mech. Div. Proceed. of ASCE, vol. 88 No. EM1, 1962.

[37] Timoshenko S. and Woinowsky-Krieger S.: "Theory of Plates and Shells", Mc Graw-Hill 2nd ed., New York 1969.

[38] Niordson F. I. (Ed.): "Proceedings of the 2nd IUTAM Symposium on the "Theory of This Shells", Springer 1969.

[39] Zienkiewicz O. C. and Cheung Y. K.: "The Finite Element Method in Structural and Continuum Mechanics" McGraw-Hill, London 1967.

[40] Pian T. H.: "Derivation of Element Stiffness Matrices" AIAA Journ. 2 (3), 1962.

[41] Jones R. E. and Strome D. R.: "Direct Stiffness Method Analysis of Shells of Revolution Utilising Curved Elements", AIAA Journal, 4 (9), 1966.

[42] Haisler W. E. and Hilbert D.: "Rigid-Body Displacements of Curved Elements in the Analysis of Shells by the Matrix-Displacement Method", AIAA Journal 5 (8) 1967.

[43] Courant R. and Hilbert D.: "Methods of Mathematical Physics, vol. II, International Publishers 1962.

[44] Connor J.: "Mixed Models for Plates" Course of Finite Element Techniques in Solid Mechanics, University of Southampton, 1970.

CONTENTS

Printed in the United States
By Bookmasters